Praise for *Road Gang*

Road Gang was a real page-turner. I finished in one evening and was rewarded with a lively, candid account of the life and times of a fine Viet Nam era soldier.

Max Toch – Army Engineers,
Republic of Vietnam 1968-70

*

If life is a road that we walk from one end to the other, the detour that many of us took during the sixties and seventies through Vietnam certainly made all of life's other little detours seem pretty inconsequential.

From the moment of arrival in-country until the phone call arriving back home, it is a path that all of us took; and proceeding along with Bo day by day was a walk down memory lane – a comfortable visit with fellow travelers back to a time that today only exists in our memory. Ironically, his tour ended in the same province and in the same district where mine started just a few days after he left. While it is true that there is no way to convey all that combat entails to someone who has not been there, this sharing of a year in harm's way resonates with those who went that route, and it is simply enough written and compelling enough to convey the frustration, boredom, occasional terror and infrequent fun in a manner that those who missed the experience can relate to and perhaps at least gain some understanding.

Wayne Marshall – Phoenix Program,
Mekong Delta, Republic of Vietnam 1970-71

*

It reads smoothly and quickly. Well done from a flowing narrative perspective....

We infantry guys always wondered what you combat engineers were doing back there anyway and now we know! I was very impressed how your experience with thermodynamics, chemistry and math resulted in actual combat construction application and solutions.... Another thing that I had not appreciated about you Engineers is that you had to understand and work with strange-sounding things like Laterite, Sheep's Foot Rollers and Melted Babbitt Bearings.... Made me appreciate being an infantryman where things were much simpler. All we really had to know was where we were and where the NVA were so we didn't end up shooting at the wrong folks!

As you said very well, Vietnam was where the action was and it seemed to be our duty to be where we were needed most.... Your work is a most commendable and thoughtful account that really brought home the Vietnam experience that we were all thrown into. You put everything in wonderful perspective!

Albert P. Burckard, Jr. – CO A Co., 2/506th Infantry, 101st Airborne Division, Republic of Vietnam 1970

*

A compelling and inspiring autobiographical tale of leadership of an engineer construction unit during the Vietnam War. The daily challenges of meeting an ambitious and flawed construction schedule with limited resources and under the continuous clouds of security and weather concerns are described in precise detail. The author's own maturation process in the role of company commander, as illustrated by priorities set, decisions made, and self-reflection regarding orders and feedback received from higher command, is told in a thorough and thoughtful manner. In the end, he achieves a high level of personal growth and obtains invaluable leadership lessons learned as a result of this challenging experience. The "Never Say Die" VMI Spirit is exemplified throughout.

Andy Yowell – Engineer Adviser, MACV Advisory Team No. 42, Binh Dinh Province, Republic of Vietnam 1969-70

*

As a Vietnam era USAF veterinarian, I did not serve "in country." As a result, I have the ultimate respect for those who did, whether or not they were in the rice paddies or in support. From "Sobering Statistics from the Vietnam War" it was noted that in 1995, 1,718,823 of those who served "in country" were still alive out of the 2,709,918 who served in Vietnam. During that same census 9, 429,958 Americans claimed to have served "in country." A pathetic statistic. This book gives an insight into what went on in Vietnam and the conditions under which Americans served their country whether they thought the Vietnam War was right or wrong. An interesting, insightful and entertaining read that I personally enjoyed.

Doug Jernigan - DVM, USAF, 1970-72

*

This account by a young officer from VMI assigned to an Army Engineer unit and plunged into the Viet Nam war is disarmingly simple in its straightforward, to-the-point narrative; but with rich detail it chronicles a triple pressure: the constant threat of an attack by the Viet Cong, overcoming the nearly impossible physical obstacles to building the road, and fighting the tension from faulty decisions from higher ups in the army chain of command. Through it all you can feel the strong bonds between Bo and his men.

This is particularly meaningful to me, since nearly 70 years ago I was a brand new 2nd Lieutenant from VMI assigned to a Combat Engineer platoon: during WW II in France and Germany, a different time, a different war, a different mission but I could feel those same powerful emotions that are so well presented here.

Brian H. M. Bowen – Platoon Leader,
10th Engineer Battalion, 3rd Infantry Division,
World War II

*

The Road Gang was a great read. It took me back to a time and place I had put far away. Like many other veterans of that war, I have stayed away from books on Viet Nam. In fact I have avoided talking about the place or my time there except on rare occasions and circumstances.... Your excellent book helped me understand many things as well as forced me to again search my feelings on the war.

I believe each one of us has a different view of that war based on our unique experience in country. This personal experience creates an unseen wall around each of us – protecting us from a return to some hurtful event or time we lived through over there. Though proud of my men, our unit, and what we did, for years I could not talk about our war without feeling a great deal of sorrow and a humble contriteness – difficult if not impossible to explain. Your book addresses this feeling....

Many of the real life situations in the book captured the poisoned social and political climate of the time as well as the struggle our soldiers and their families experienced as a result. It is impossible to understand the internal conflict each soldier struggled with; fighting a war while trying to explain and justify their actions to those they loved and depended on for a lifeline of support at home. One telling event in the book is that of a soldier/officer travelling to Hawaii on R&R to see his wife, only to find out when he arrived she had decided not to come, leaving him alone, abandoned by his home and family.

The Road Gang captures the pressure of deadlines, mission creep, increased work load demands, political intrigue, and provides an excellent example of making "Bricks without Straw," or getting the job done without the means necessary to accomplish the mission while fighting an ever present enemy. This book makes you look at real life and death issues, where all management and leadership challenges reach a climax dictated by an uncompromising deadline fixed by mother nature. There can be no time extensions....

This exercise has forced an old soldier to look back and reconsider some of my past actions. My relook has made me realize luck is more of a factor in life than I ever wanted to admit....

James N. Joyner, Jr. – Commanding Officer
E Company, 1/5th Cavalry, First Air Cavalry
Division, Republic of Vietnam, 1969-70

Published Works by the Author:

Books

Road Gang: A Memoir of Engineer Service in Vietnam (2014)

Empire of the Owls: Reflections on the North's War Against Southern Secession (2013)

Featured Articles:

"Reflections on Energy, Technology and the Lost Cause," *Confederate Veteran Magazine* (Nov./Dec. 2013)

"Seafaring on the Waterway," *Chesapeake Bay Magazine* (Nov. 1994)

"Winter Hurricane," *Chesapeake Bay Magazine* (Mar. 1994)

ROAD GANG

~

A Memoir of Engineer Service in Vietnam

By

H. V. Traywick, Jr.

And through some mooned Valhalla there will pass

Battalions and battalions, scarred from hell;

The unreturning army that was youth;

The legions who have suffered and are dust.

- Siegfried Sassoon, "Prelude: The Troops" (1)

First Printing

Cataloging-in-publishing data for this book available from The Library of Congress

ISBN: 978-0-9903687-8-6

H. V. Traywick, Jr., Author

For information write:
Dementi Milestone Publishing, Inc.
1530 Oak Grove Drive
Manakin-Sabot, VA 23103
dementi@aol.com
www.dementimilestonepublishing.com

Page layout and Cover Design by Dianne Dementi

Printed in the USA

Affectionately dedicated to

the men of

THE ROAD GANG

D Company, 34th Engineer Battalion,

20th Engineer Brigade

Republic of Vietnam

1969 - 1970

Capt. Traywick and Sgt. Mendoza at the 1st
Laterite Pit

34th Engineer Battalion
Highway 1A/2A Upgrade Project – Republic of Vietnam
November 1969 – May 1970

The Road Gang at the Lay-Down Site – Highway 2A

TABLE OF CONTENTS

LIST OF ILLUSTRATIONS

FOREWORD

By

Phil Gioia

Though the iconic image of the Vietnam war was the helicopter, in reality almost everything needed to fight the war moved by truck. All classes of supply: ammunition, rations, fuel, medical, communications, and construction, had to be hauled by truck convoy, from ports and depots near the coasts to the fighting divisions in the field. Accordingly, reliable heavy-duty roads were needed, most important of which were the main supply routes, or MSR's, leading to the base areas of the combat divisions.

The unsung heroes of maintaining the MSR's, and where necessary of building them, were the road construction units of the Engineers. 'Road Gang', Bo Traywick's tale of his year in the field as an Engineer company commander, transcends the simple memoir. He brings to life the sights, sounds and smells of the terrain and weather, the clanking, dusty machinery of his road construction company, and the stretch of the critical MSR it was assigned to build. His self-deprecating style does not conceal his professional soldier's dedication to the mission and his men, his deer-hunter's wariness of the enemy, and his race to complete the road before the onset of the monsoon rains.

Traywick tells his story with the *patois*, common sense, and country instinct of his native Virginia. Though he pokes fun at himself for being a 'four-year private' as a cadet at Virginia Military Institute, as his narrative proves he emerged from VMI a savvy civil engineer, with a rare talent for improvisation whenever the mission required it.

More than a war story, 'Road Gang' is a remarkable tutorial on how roads are built, and a sort of 'Poor Richard's Almanack' of field engineering, with hard-learned maxims such as 'water is the enemy of a road'; 'if you don't take the time to maintain equipment when you have it, the equipment will take the time when it wants to.'; and 'a washed-out culvert has the same effect as blowing a bridge'.

Appearing in 'Road Gang' are colorful characters like Colonel George S. Patton, the swashbuckling armored cavalry officer and son of the famed World War II armor commander; a heavily-armed Army Chaplain, whose uncommon sense of practical tactics presents Traywick with a solution wrapped in a moral dilemma; an Engineer Major General and VMI *alumnus* who leaves a lasting impression on Traywick, before flying off to his own fate and destiny; a *cadre* of professional NCO's, and the ordinary citizen-soldiers swept up by the Vietnam-era draft, who provided the core of most in-country units.

Those of us who fought in the field as combat leaders in Vietnam did so with little knowledge or appreciation of what it took to enable timely delivery of the mission-critical supplies we needed. Not until I read 'Road Gang' did I learn the man who maintained the MSR to our combat division in 1969-70 was my VMI classmate and 'Brother Rat,' Bo Traywick. He is the man you want on your side, in a tight corner.

Phil Gioia
CO, Company A, 1st Battalion 5th Cavalry
1st Cavalry Division (Airmobile)

Phil Gioia grew up as an "Army Brat." Graduated from Virginia Military Institute in 1967, he was a Ranger, Master Parachutist and Pathfinder. He served as an Infantry officer in Command assignments in Airborne and Airmobile units in the United States and Vietnam. During two combat tours, he received two awards each of the Silver Star and the Purple Heart. He is a businessman in California and writes on topics of military history.

INTRODUCTION

In the summer of 1969 young American men were being sent to Vietnam to fight in a war that many felt had lost its meaning. The "Vietnamization" of the war had just begun and US troop units were being withdrawn and replaced by units of the Army of the Republic of Vietnam (ARVN.) The author of Road Gang was one of those arriving "in country" to serve a year's tour of duty during this period of transition. The heart of his narrative is an account of his service as the commander of an Army Engineer company engaged in the construction of a section of the Main Supply Route (MSR) to the 1st Air Cavalry Division, but his other experiences are touched on as well - including service with the 82nd Airborne Division at Phu Loi, and duty as an Engineer reconnaissance officer on the Cambodian border.

This work offers a glimpse of the Army Engineers and some of the work they did in the Vietnam conflict. The focus is on the officers and men of "The Road Gang" - D Company, 34th Engineer Battalion, 20th Engineer Brigade - and the road they built, but this straightforward narrative transcends the simple memoir. Behind it lies a portrait of the poisoned political atmosphere of the times, the mortality of an army, and the end of an era.

H. V. Traywick, Jr.
Richmond, Virginia
October 2014

ACKNOWLEDGMENTS

I would like to thank the following people for helping to make this work possible:

First, I would like to thank Wayne and Dianne Dementi, and Dementi Milestone Publishing, for taking on another of my projects. It is a great pleasure to be associated with you, and I am sincerely grateful to you for your efforts in bringing out this work.

I am particularly grateful to my VMI classmate, Phil Gioia, for his most generous comments in the Foreword to this work. Little did I know at the time of my service on the Main Supply Route (MSR) to the 1st Cav, that my Brother Rat was up there at the far end of it on the "Thin Red Line," mixing it up with the NVA!

I also wish to offer my profound gratitude to my Brother Rats who served in Vietnam and who kindly consented to review this work and offer their generous endorsements: my Brother Rat and Ranger Buddy Max Toch, who learned how to freeze and starve with me in the mountains of North Georgia; Wayne Marshall, who has "seen the elephant" under the towering skies and the golden green marshlands of the Mekong Delta; Andy Yowell, who studied Civil Engineering with me at VMI and who ended up "in country" serving with the Engineers at the same time I was there; Bert Burckard, who fought with the "Screaming Eagles" for firebase Ripcord; and Jim Joyner, who was also on the "thin red line" at the far end of the 1st Cav's supply line. I'm glad you all made it back.

I would like to thank Dr. Doug Jernigan, a Vietnam-era veteran of the Air Force, for his generous endorsement and for his service during those trying times. We serve where we are sent and try to do our best when we get there.

I would like to sincerely thank Col. Brian Bowen - Combat Engineer Platoon Leader in Europe during the Second World War, recipient of the Silver Star and a friend of my father - for his warm endorsement of these modest efforts of mine. I never saw what you or my father had to see.

I would like to thank my two lovely cousins, Maureen Luby Cummins and Michele Luby Eldredge, for their tireless proof-reading, their comments, their suggestions, their interest in the project, and mostly for their encouragement. You have given me inspiration, and I am very grateful to both of you.

I want to thank my sister, Robin Traywick Williams, and my Niece, Katherine Bolling Williams, for their moral and technical support. I know that I have imposed upon you – if I have not nearly pestered you to death – but I want to thank you very much for your patience and your help.

I especially want to thank my loving family – my Father, my Mother, my sister Robin, and my brother Cris – for their support and their prayers while I was away living these things. You gave me the best Welcome Home anyone could ask for.

Heartfelt thanks go to Bob Hope, to Johnny Grant, and to two young ladies from California who came over to Vietnam when they did not have to and gave a touch of caring to some who needed it pretty badly. There are many who will always hold you in the warmest regard, and I am one of them.

And last, but not least, I would like to thank every man who served in D Company, 34th Engineer Battalion, for the support that you gave me. You did your duty and did it well, and I will never forget you.

PROLOGUE ~ A LETTER HOME

31 March 2004

Dear Mom,

Last night you told me I never said much about Vietnam. I guess not - not because I have any "deep psychological scars," etc., etc., but because I have felt that I had nothing much in the way of war stories to tell.

I have heard a few, though, and I know that many of those men who were there went through pure hell; and many will live in hell from now on with maimed bodies and souls. That is no different from any other war. But with this one you had the nagging sense that it was all for – what?

We knew about the "Communist Bloc" and the "Iron Curtain" on the one hand, and about the "Western Democracies" and the "Free World" on the other, of course. We also knew about the "Third World" where the "proxy wars" were being fought after the Atom Bomb made for a paradigm shift into "limited wars." We had heard something about "The Domino Theory" and how we had to draw the line somewhere against Communism; how we had to back whomever "our guy" was – no matter how unlovely - in whatever Third World country, so it wouldn't fall to the Communists. "Either fight 'em there in the jungle, or fight 'em here in the streets," that was the buzz back then - along with the antiphonal chant, by some, of "Better Red than dead!"

Nonetheless, in the revolutionary sixties it was hard to get up a crowd to fight and die in order to keep some little country nobody had ever heard of from going Communist - or Buddhist, or Shinto, or Shiite, or whatever else they wanted to be. It was their country, man. “Stop the Domino Effect!” might have been the battle-cry of the Cold War, but it was hardly the sort of thing that would have inspired fifteen thousand boys to fix bayonets and charge up Cemetery Hill into a solid storm of shot and shell, or to huddle in churning landing crafts under a continuous barrage of artillery fire in the pre-dawn darkness, waiting to “hit the beach” on Normandy or Iwo Jima. And it didn’t inspire an overabundance of us, either. But we went anyhow, and those of us who served honorably and made it back, I think, are proud to have served. Plus, it was the only war we had, and – Duty notwithstanding - if you wanted to know how you were going to do, you had to go and find out.

Our fathers had gone to their war and – those who survived – had come back without a bunch of fuss. But in their case the whole country was behind them and pulling for their efforts in that last total war. Vietnam – and all limited wars since - was different in this respect. An ambush patrol was the same for a “grunt” in Vietnam as it had been for a “grunt” in the Solomons – up close and personal. But on the home front, things were different. Limited wars generate limited participation. To those who had sons or husbands in Vietnam, the TV footage on the 6 o’clock news meant something entirely different than it did to those who didn’t. The soldiers knew it, too, and they knew they weren’t going to be coming home to any ticker-tape parades on Wall Street – if they came home at all. Thus there appeared the phenomenon of the ragged, outcast, and forgotten vet alone in the misty morning stillness, weeping at The Wall for something lost forever – and then later in the day, maybe, some not-so-poignant others showing up to sob on each other's shoulders for the benefit of the tourists and the TV cameras.

But the men who suffered the greatest losses are the ones least likely to be among those trying to elicit sympathy from the public, for they know that it was the public, that it was "We, the people," the halfhearted people and their representatives – and perhaps "The Masters of War," as Bob Dylan called them - who sent them over there in the first place, and then – at least in the case of those hiding behind protest signs and spurious principles in order to mask their weaker characters - spat upon them when they got back home. These men might have lost a lot in the interim, but not their self-respect, so they keep their mouth shut as they try to get on with things. They call it being "back in the World," for they have come from a place far away.

They don't wear their sense of alienation on their sleeves, but they have a particular reserve, a part of them inside that they will seldom unlock for anyone who isn't "one of them." They are different; and if *you* can't tell them apart from others, it is a tribute to their rectitude. But *they* can unerringly pick each other out in a crowd, for they have a certain highly developed and dearly bought perspective on things. They know what is important, because they have seen men die for things that are not. They are not swayed by demagoguery, for they know the cost of political hype and have had to pay the price in blood. They do not worship the "heroes" manufactured by the entertainment or the sports industry, for they know the difference between a hero and a celebrity - and they know the worth of each. They are not moved by the hustlers in the market place, for they can spot bull feathers from fifty miles away with the naked eye. Some call it cynicism. They call it a survival trait. It is of Darwinian proportions, and they have come by it honestly.

And all for - what?

I guess all of us asked ourselves that question at some point. Speaking only for myself, I remember wrestling with it, with my answers running the gamut from "so that maybe my brother or my son won't have to do this," to considering myself as being an underpaid mercenary. I remember Dad telling me before I left that it was a dirty political war, but to do my Duty, keep my head down, and get my men back home. So in the end, I reckon, most of us determined that we were there for our brothers-in-arms. That was reason enough.

But at any rate, thirty-four years gives one a better perspective on things, so with the wisdom gained during that interval, I hereby submit this humble account of an "unrenowned warrior" to you, my dear Mother – who cried over the telephone when I called you that night from a bar in New Jersey to tell you I was home….

Cadet Traywick

VMI Class of 1967

CHAPTER I

INTO THE BREACH

I remember leaving Richmond, Virginia, for the Republic of South Vietnam on that late August day in 1969. My family was at the airport to see me off. I was in my Army khakis, with Captain's bars on one collar and an Engineer castle on the other, a Ranger patch on my shoulder, and Airborne wings on my chest. I was a Professional. Some would have accused me of being a "Lifer," but I had never intended on making a career out of service in the military. I was indeed bound for San Francisco *en route* to Southeast Asia that day, but I confess that I took a little detour *via* Nantucket for a little farewell fling, which, by the time I did get to California, had pretty much set me back to nearly broke - and did indeed do so before I got to the Presidio for transit out of the country. My Father - bless his heart - sent me enough to get me to where I was supposed to have been going in the first place, and never said anything more about it.

That was a right lonesome time at the end. I remember my Father saying that when he was in San

Francisco to ship out for Burma in World War II, no one in uniform was allowed to buy himself a drink – the good citizens of that city (those who were not able for reasons of age or health to get themselves into uniform) crowded the bars and picked up all the tabs. Things had changed. There was some time to kill before the shuttle to the Presidio was to come along and pick me up, and I remember going to a theater by myself while I waited for it. The movie was *Castle Keep* with Burt Lancaster, a fanciful film about WW II. I remember watching the movie by myself with my bag between my feet on my last day among the States. I didn't want to think forever.

I processed at the Presidio and bunked on a cot for the night. Next day we flew out on Braniff Airlines. Our flight stopped over in Hawaii for fuel and we got off the plane and stretched our legs in the airport, but were soon gone again. We probably made another fueling stop somewhere – maybe in the Philippines – but I don't remember. Coming in over South Vietnam at night you could look out of the plane window and see the flashes of artillery fire – probably H&I (Harassing and Interdicting) – somewhere out in the countryside. The pilot called our attention to it over the intercom.

We landed at Tan Son Nhut Airbase, near Saigon, in the early morning. The first thing that we did was to go through a building consisting of planks and chicken wire to turn in all of our cash for "MPC," or military scrip. This MPC was to be used for all of our financial transactions while "in country" and had been devised to combat black-marketeering. It was illegal to use with the Vietnamese and their economy. To deal with them we had to exchange our MPC with the Army Finance people for Vietnamese *piasters*, or "p," as it was often called. This was not always done, however, because MPC bought more "bang for the buck" than did the local currency. The Vietnamese would deal in it eagerly but it was like a hot potato. Periodically the U. S. would change it without warning and declare the old scrip illegal tender. We would then go and exchange our old MPC for the new; but *papa-san*, the entrepreneur of the black market, would be left holding a bag full of worthless scrip.

I exchanged my money, which I remember totaled precisely one dollar and fifty-three cents. You could not exchange pennies, all of which went to a Vietnamese orphan relief fund, so I ended up with $1.50 in MPC in my pocket and 365 days in country to go. I had this philosophy that the further in debt I was upon arrival, the more fervent would be the prayers of my creditors for my survival. Acting upon this premise, I had engaged the solicitude of BankAmericard with four or five days in Nantucket, a night in San Francisco, and a diamond ring. Whether or not this theory had any effect on my subsequent health and well-being, there is no way of knowing. I do know that the solicitude of my creditors was of a hardihood that eclipsed that of the young lady upon whom I had bestowed the aforementioned attentions. I suspected this after a few weeks in country, when one day I received seven perfumed and tear-stained letters, and the next day I never heard from her again for the rest of my life. But BankAmericard was more constant, and I continued to hear from them on a – one might say - painfully regular basis. I know that I didn't see a payday worthy of the name for the next eight months - Captain's pay, jump pay, and combat pay notwithstanding.

We boarded the buses at the airfield and were taken through the city somewhere to the replacement compound. There was wire mesh on the outside of the bus windows, so that nobody could pop a grenade into the bus. We spent the night at the compound where our assignments were to be posted. Everybody was hoping not to be sent "up-country."

I remember seeing Maj. Ken Dickerson's name on the "up-country" list for the 173rd Airborne Brigade. This was another one of his tours. He had been a tactical officer at the Virginia Military Institute and taught our Infantry ROTC. He was a Captain back then and an Airborne Ranger with a Combat Infantryman's Badge. He was a Master Parachutist, or "Master Blaster," which meant he wore a star with a wreath on his jump wings signifying 36 – or however many – jumps, plus qualifications as Jumpmaster and Rigger. He once told us about jumping in a "daisy chain" from the ramp of a C-130, where everyone – evidently bored with plain old jumps - hooks up his static line to the man behind him and only the last man in the stick hooks up to the wire. Thus the first man out is the last one to have his chute deployed. Fun for the feeble-minded.

It was Capt. Dickerson, I'm sure, to whom I owed my designation as a Distinguished Military Graduate from VMI, and thus my Regular Commission. It was certainly not because of my academic standing, or my standing in the Corps of Cadets. First Classmen didn't attend the Spring Field Training Exercises, but there was an opportunity at FTX during my First Class year to work off some penalty tours (instead of marching them off with the "Wednesday and Saturday Afternoon Hiking and Gun Club"). Some help was needed to set up some of the courses for FTX beforehand, and I volunteered to go out one afternoon on Penalty Tour Detail to help rig a rope bridge across the river.

Capt. Dickerson was in charge and he showed us how to mount the rubber assault boat so we could tow the first line for the bridge over to the other side. We paddled across the rushing river towards the far shore, which was nothing but a hill that rose straight up out of the river. When we got there the man in front of me grabbed a bush sticking out of the side of the hill and the current pulled the boat out from under him. He went into the water and was swept down the river through the rapids. I jumped out after him and grabbed him by his shirt collar, and we both were swept away together. We fetched up on a dry rock and he said thanks, but he would have been *much* better off if I hadn't tried to drown him!

After this little "Chinese fire drill" was over, I was driving the deuce-and-a-half truck back to the VMI Post with Capt. Dickerson riding in the cab with me. I remember him asking me what I wanted to do after I graduated and took my commission. I told him I'd like to go to Ranger school and all of that, and maybe Special Forces. I remember him saying that Ranger school would be all I'd want. It was. Anyway, I never thought about that conversation again until the day a few months later when they announced the Distinguished Military Graduates. My name was on the list. Of all the surprised people, no one was more so than I was. The only way I could think of that it could have happened was for it to have been placed there at the recommendation of Capt. Dickerson. I never found out.

A DMG got offered a Regular Commission - instead of the commission in the Reserves that everyone else was to get - and he got to pick his branch of service. I was a Civil Engineering major, so I accepted the commission and chose the Army Corps of Engineers. I remembered that General Lee had taken his commission in the Engineers.

Upon graduation I had to go through ROTC Summer Camp at Indiantown Gap, Pennsylvania. I didn't make it to Summer Camp with my own class the previous year when I was supposed to, because I had to go to summer school instead and give Thermodynamics and Electrical Engineering another try.

After Summer Camp I was commissioned a Second Lieutenant in the U.S. Army. This was somewhat of a change in status for me, for, although I had once at the end of my Rat year been suggested as a possible candidate for Lance Corporal – amidst gales of laughter by the Third Classmen who lived in the room in barracks below ours – the highest rank I had ever actually held in the Corps of Cadets was "High Private in the Rear Ranks." But I was not daunted by the transition, because although my innate sense of modesty prevents me from suggesting that a First Class Private in the VMI Cadet Corps is the equivalent of a Second Lieutenant in the United States Army, my love of the truth compels me to call attention to the fact that only a mere step separated the two.

I was sent to Fort Belvoir for Engineer Officer Basic Training, where we learned about bridge-building, demolitions, calling in fire support, mine warfare, and other basic aspects of military engineering. That lasted a couple of months, and then I was sent to Fort Benning to the Ranger School. Most of us there were young officers, but the instructors and lane graders were NCO's, so for the three months that we were there none of us wore our rank, and any authority ordinarily accorded such rank was put on "hold" for the duration of the training.

Our instructors were some of the best and the toughest men in the Army and we all respected them as such. They were members of the 75th Ranger Regiment, and we were proud to be Ranger Candidates in the Training Battalion, 75th Rangers. Their mission was to teach us to lead men effectively under the harshest conditions that could be devised short of actual combat, and they had developed their methods to an *excruciating* perfection. My "Brother Rat" Jim Joyner, who served an extended tour as the commander of an Infantry company in the 1st Air Cavalry Division, said he never went through anything like it in his eighteen months in Vietnam.

We spent a month at Fort Benning, living on short rations and little sleep, going on forced marches, running obstacle courses and confidence courses, learning hand-to-hand combat, (just enough, it was said, to get your butt kicked in a bar room,) and learning how to run combat and recon patrols. We were taught always to think "security, security, security," and that paying attention to the details will save your life, and the lives of your men. *Details* and *security* were drilled into our heads day and night, and throughout all of our training. They were maybe the best lessons we ever learned anywhere.

After Fort Benning, we were sent to Camp Dahlonega, in the mountains of north Georgia, where we learned rappelling, mountain-climbing, and river-crossing techniques, along with land navigation and long-range patrolling. We endured hunger, exhaustion, and hypothermia, while humping machine guns, ammo and gear up and down mountains all day and all night, often in a freezing drizzle, (no ponchos allowed – they rustle, which impairs your hearing and gives your position away), dodging aggressors, and securing helicopter landing zones (LZs) for our meager resupply, (one C-ration per day – *if* you were lucky!) And just about the time you couldn't think of anything at all but how utterly miserable you were, and with no hope of any relief in any foreseeable future, the lane grader would pronounce the patrol leader KIA, and tell *you* to take charge of the patrol and continue on with the mission!

In this mountain phase of Ranger School, when we weren't on a patrol we lived in little cabins heated by wood stoves. It was the only time I ever remember being warm. We learned river-crossing techniques, including how to tie knots and handle rope properly, and how to build a rope bridge. We learned how to cross a river on one rope, as well as how to construct a three-rope foot-bridge. The problem was that one hearty soul had to get that first line across the river to secure it on the other side. November in the mountains is not the time one ordinarily would wish to go "skinny-dipping."

We also learned mountaineering, how to drive *petons* and rig climbing ropes, how to belay, and how to rappel. To rappel, you first take a short piece of line and tie it around your waist and between your legs in a certain fashion to make a seat. Then you take a snap-link and snap it into the front of this seat by your groin. You face the anchor point of the rappelling rope, pull a bight of the rope towards you, and snap this turn into the snap-link. You hold the standing end of the rope in one hand and lead the running end around to the small of your back with the other to be used to control your descent. You then back yourself over the side of the cliff, holding your body perpendicular to it, and either walk down backwards or push out in a series of self-controlled drops, alternately slacking and tightening your brake until you reach the bottom.

At one phase of our mountaineering training we had to rappel off a cliff with our "Ranger Buddy" (Rangers *always* operate in pairs – I hope that information is not classified) on our back. My Ranger buddy was Max Toch, one of my Brother Rats from VMI. Max had played end on the VMI football team and weighed around 220 pounds or more, and stood about 6'3" or so. I was 5'9" at about 140 pounds. Max went first, with me on his back, and we did fine. Then I had to go with him on my back. Max had a leg about as big around as a 55 gallon drum and I couldn't get my arm far enough around to get the brakes to work very well. By the time we got about halfway down we were accelerating at a pretty good rate and my gloves were starting to smoke. We hit the ground like a ton of bricks flat on our backs. Luckily I landed on Max, so I was OK.

I later read in the Field Manual where when you rappel with a man on your back you are supposed to take two turns on your snap link. I must have missed that part of the lesson. Max went on to glory in Vietnam, serving with the Combat Engineers of the 101st Airborne, and earning the Vietnamese Cross of Gallantry. He is still speaking to me, by the way.

The last phase of the school consisted of a month in Florida for the "Jungle" phase. Here we learned to handle rubber assault boats and to navigate with them in the streams and swamps. We endured more hunger and exhaustion and long range patrols, although thankfully it was not as cold on the Gulf Coast as it had been in the mountains. On the patrols we were not allowed to take any extra rations. I had secured one of those flat cans of smoked oysters before we left on a ten-day patrol and was munching them clandestinely when the lane grader dropped back towards my position. I quickly stuck the can into my fatigue pants pocket until the way was clear to finish my repast. But the oil ran out of the can and soaked my pants-leg, and I slogged along for the remainder of the ten days smelling like a fish cannery. I have never liked smoked oysters since, whether I was starving or not.

The best Christmas present I ever received was the award of my Ranger Tab upon the successful completion of the training. This I could wear on the left shoulder of my uniform. Less than half of our class graduated, although some who had been taken out of the mountains with bronchial pneumonia were recycled through the school at a later date and were successful.

After Ranger training I was assigned to Fort Stewart, Georgia, with the 575th Engineer Battalion, where I served in various capacities ranging from Platoon Leader to Company Commander. The battalion's mission was to provide engineer support for the helicopter- and tank-gunnery ranges there, as well as general construction support for Fort Stewart, and the neighboring Hunter Army Airfield at Savannah. One of our projects, I remember, was the construction of a perimeter road around one of the tank gunnery ranges. Another one was the construction of golf-cart paths for the golf course.

Fort Stewart had been President Kennedy's Headquarters during the Cuban Missile Crisis. There were rotary-wing and fixed-wing flight schools there, and the place was full of jaunty pilots-in-training. We stodgy old "dirt-daubers" didn't have quite the same flair around the Officers Club as they did, I can tell you. But I ended up sharing the rent with four other officers at a house in Hinesville that had a swimming pool, so we found life there to be not entirely unendurable. At one point I owned a motorcycle, purchased from one of my housemates after I had a little mishap with it trying to jump a ditch behind the officer's club, etc., etc. (There were "extenuating and mitigating circumstances" involved.)

On our FTX I was the "Aggressor Chieftain." I was given a squad of men and we had fun setting up ambushes and conducting other scenarios in the script assigned to us. I remember setting up onc night ambush using a series of quarter-pound blocks of TNT strung together with “det” cord, which we fired electrically when the convoy passed. Then we shot our M-14’s up into the air - loaded with blanks, of course - and generally made a bunch of racket before we slipped back into the woods. We were given times and places to spring ambushes, make assaults, and so forth. Of course that got to be a little prosaic, so near the end of FTX we decided to improve on the script a little bit and make a simulated “suicide raid” on the battalion encampment one evening to liven things up. Our little impromptu adventure involved a deuce-and-a-half truck and some artillery simulators, which we distributed liberally as we roared through. The Colonel was not as amused as we were and the little episode caused me to get fired as Aggressor Chieftain, but nowadays, I suppose, the inclusion of such drills is probably Standard Operating Procedure (SOP) when units go on FTX. We were just a little ahead of our time.

In 1968 I went Temporary Duty (TDY) to Panama for the Jungle Operations Course. By then I had been promoted to First Lieutenant. I went with a Captain in our unit who was a naturalized U.S. citizen from Sweden. We landed at night in the Republic, on the Pacific side, at Panama City. From there we had to go through a checkpoint to enter the Canal Zone. We then caught a train across the Isthmus to Fort Sherman. The old coaches had ceiling fans to stir the balmy tropical air, but they were not needed for comfort that night. We stood on the platform on the back of the last car, and as we rocked along with the clicking of the rails we watched the jungle receding down the tracks into the darkness. One could almost expect at any moment to turn back into the coach and find Humphrey Bogart in a white linen suit lighting a cigarette for Lauren Bacall.

Jungle School was like "Ranger School Lite." It was two weeks long, but we ran some patrols in the jungle, ate some rats and snakes, and did a little rappelling. One thing we did there that we did not do at Ranger School was to go through an Escape and Evasion Course (E&E.) They told us of the fun and games that the aggressors had planned for us if they caught us - and we did not want to be caught! Like hanging, the prospect focused one's attention, and I did not get caught.

I learned one lesson there that I have never forgotten, which is, that the jungle is not your enemy; neither is it your friend. The jungle – like the sea - is indifferent, and that indifference is monumental. It helps to keep man's own little endeavors - like war - in perspective.

In Panama, down by the equator, the "dawn comes up like thunder" - both figuratively and literally. Many mornings a torrential downpour would come over, rinsing the sky a cobalt-blue and leaving the clouds rolling up into the heavens as the squall passed over the anchorage that lay off Fort Sherman. Ships would anchor there to await transit of the Gatun Locks, which would lift them into the Panama Canal for their passage across to the Pacific. Out in the jungle you could hear them sounding their horns. There was one in particular that I remember hearing while we were out on a patrol. It was pitched so low that you could almost feel it more than you could hear it, and it reverberated like the trump of doom through the jungle canopy along the Chagris River.

While at Fort Stewart I had been trying to get into the Parachute School at Fort Benning, but it was a busy time with the war going on full blast and I think the Infantry was getting all the slots. I finally got in just before I shipped over to Vietnam. It was a three-week school. I got a room in the Bachelor Officers Quarters (BOQ). We didn't have to live in the mud and freeze to death, and you could go get a beer at the end of the day.

The first week was called "Ground Week." We spent most of that week drilling in Parachute Landing Falls (PLF's.) We did them off a platform into the sawdust – forwards, backwards, sideways – until you could do them in your sleep: toe, ankle, calf, thigh, then roll onto the side and back. Then they had a pulley on a sloping wire that you would hang from that gave you a simulated lateral drift before you turned it loose and hit and did your PLF. Next they rigged you up in a parachute harness and hoisted you off the ground so you could practice "pulling a slip" in the risers, to cut your lateral drift on the Drop Zone. Hanging in harness for an hour of that was guaranteed to get your britches in a bunch.

But repetition is the key to training. I remember, many years later, sliding off the tin roof at my house in Deltaville when the rig I had made gave way. No problem. I just executed a perfect five-point PLF when I hit the ground. It would have been fine had it not been for the big rocks that bordered the flower bed, which broke a couple of ribs. But it proved the point that the Army teaches, which is, that when you are in a stressful situation and don't have time to think, you do automatically what you have been trained to do.

The other little thing that they wanted you to get used to was the idea of jumping out of a perfectly good airplane. To get you inured to that idea, they rig you up in a parachute harness with a bungee cord attached in place of a static line, and march you up into the 34-foot tower. Thirty four feet is psychologically the most daunting height to deal with because it is as high up as you go before you start loosing your sense of depth perception. The tower has a cable outside of the mock door up there that you hook your bungee cord onto. The cable runs down to a berm about 75 yards away. You hook up and jump, assuming the correct exit posture (which we also practiced to the command of "Hit it!" about six thousand times a day,) and ride the cable down to the berm. There the man who went out before you awaits to catch you and help you un-rig.

In addition to training, we ran. We ran between drills and we ran while we were resting. We ran wearing T-shirts, fatigue pants and jump boots. We ran in formation around the Post doing the "Airborne shuffle" and singing about "Jody" to keep the cadence:

> Jody, Jody's got big feet –
> Stretch from here to Market Street!
> Ain't no use in going home –
> Jody's got my gal and gone!
> Sound off! etc. etc…..

"Jody" is the accommodating friend who keeps your girl from being lonesome while you are away. This, in turn, assists in your development as a fierce warrior ready to kill somebody. Thus the benefit that Jody accords is twofold.

We thought we had been doing some running, when these two Navy SEALs began complaining that if they didn't get out there and do some running they were going to get completely out of shape before they got out of jump school. So in the evenings, while we were switching our exercise regimen to "16 ounce curls," utilizing our favorite beverage for the weight-training program, these two SEALs would go out and do some serious running. Of all the Special Ops people in all the services, I think the Navy SEALs are among the toughest of the tough.

The second week of jump school was called "Tower Week." This was when you were parachute-dropped from the two-hundred-foot tower. Your chute is fitted into a frame that holds it open while it lifts you to the top of the tower, and releases you when you get up there. There are two towers, each with four arms. Only three arms can be used at any one time, however, because of the wind. Once while we were there the wind shifted and blew a man into the tower, where he was hung up. The good old NCO's, of course, had a plan – as always - and they rigged up some sort of cable whereby one of them was hoisted up and pulled over to the man where he could get hold of him and get him back down. No sweat. But the trainee, I'm sure, still has his fingerprints somewhere up there in the steel.

The last week was "Jump Week," when we cut to the chase. You knew it had to come sometime. But by the time you sat around on the tarmac in the hot Georgia sun for an hour or two all rigged out in your main and your reserve, then loaded up and rattled off down the runway for takeoff in a rickety propeller plane, then roared and rattled over to the Drop Zone with that little voice inside of you asking all the while what you are doing this for, by the time the green light goes on you'd jump whether you had a parachute on or not, just to get some relief.

But the next thing you know, you are standing in the door with your faith in the parachute riggers absolute, and you "hit it" just like you are supposed to do, and you count to ten and check your canopy just like you are supposed to do, and it has deployed just like it is supposed to do, and from some crazy sideways upside down point of view you see the tail of the plane moving away from you dreamlike in this great silent sky, where little paratroopers are plopping out of the side of it, and chutes are blossoming and drifting away like dandelion seeds in the summer sun.

Below there is this mosaic landscape that you look at as though you are suspended over some enormous patchwork quilt that stretches out to the hazy ends of the earth. In the strange silence you ponder this quilt for a long time, until you begin to perceive some depth and definition, and realize that it is getting closer. Next you notice that there is a little bit of drift that you are going to have to deal with in a minute if you don't want to get knocked onto your butt when you get to the ground. You grab your two windward risers and pull a slip, spilling a little air from the lee side of your canopy to cant it into the wind. You can't do much steering with a T-10, but you try to stop your lateral drift as much as you can. Then - with your legs together and the toes of your boots pointed towards the ground, with your chin tucked and your elbows in tight as you are pulling down on the risers - you hit and roll, executing a perfect five-point parachute landing fall. Quickly you are on your feet and onto your chute, collapsing it before it can drag you into a stump or a rock and break your neck. Then you bundle it up and head toward the trucks, which are waiting to take you back to the airstrip so you can do it all again. It is as good a rush as you could ever want in this world.

Old troopers called those of us fresh out of jump school "cherries," with only five jumps to our credit. But I didn't consider myself a "cherry," for I had made five and a half. On our last jump I was next-to-last man in the stick. We were making the jump from a C-141, which was a cargo jet. When we got close to the Drop Zone we took the commands "Stand up!" "Hook up!" "Check static lines!" and "Stand in the door!" Everybody was all psyched up, and when the light by the door turned green the Jumpmaster gave us the "Go!-Go!-Go!" When I got to the door the light turned red, signifying that we had passed over the DZ, but I went anyway. But the Jumpmaster grabbed me by the seat of the pants and hauled me back aboard. That gave me the "half-jump."

(You ought to get *some* credit for going out of the door. Getting to the ground is the easy part!) The plane went around, made another pass, the last two of us went out, and that gave me five and a half.

After jump school and a furlough home, I was on my way to the wars. By the time I got to the replacement depot in Vietnam I felt I was as ready for what lay ahead as Uncle Sam could make me. The next morning, dressed in jungle fatigues and jungle boots – but wearing no underwear and no socks, to cut down on the prospects of jungle rot and other assorted fungi – I got my unit assignment. It was not "up country." It was at Phu Loi. I was to report to the 3rd Brigade of the legendary 82nd Airborne Division!

Ordinance Disposal with Demolitions

CHAPTER II

DEATH FROM ABOVE

The 82nd Airborne Division had one brigade in Vietnam. The remainder of the division was at Fort Bragg, where it served as a rapid deployment strike force, ready on short notice to go anywhere in the world. It's recent deployments, however, were to quell race riots in the U.S. While Northern "Freedom Riders" were busy down South delivering tutorials on proper race relations to the benighted Southerners, Blacks up North were burning their cities down.

But the 3rd Brigade of the 82nd Airborne was in Vietnam tending to other matters. At first it was up near Hue fighting the North Vietnamese Army (NVA), and then, in October of 1968, the unit was given an area of operations (AO) down in III Corps in the "Iron Triangle," north of Saigon. Here it began using the tactics of guerrilla warfare – conducting sweep operations during the day and ambush patrols at night – against the Viet Cong (VC). The 3rd Brigade was also involved in civic action and pacification programs among the local populace, trying to "win the hearts and minds of the people."

My first assignment with the paratroopers was to go through the brigade's week-long Combat Proficiency Training. This course was intended to provide each replacement to the brigade with a refresher of combat drills and familiarization with weapons. We threw hand grenades, fired the M-16 rifle, the M-79 grenade launcher, the M-60 machine gun, and practiced fire-and-maneuver tactics and other squad combat drills – all the basics.

It was here that I was introduced to the newly issued M-16 rifle, which - compared with the M-14 that I had been used to at Fort Stewart and Fort Benning, and the M-1 at VMI – seemed to me like a little plastic toy. I never cared for it too much and neither did some others. Back then the thing had a bad reputation of jamming at the most embarrassing times. The problem, I have heard since, was bad powder in the ammunition. The rifle was designed on a certain brand of powder, but before some key Senator would vote his approval for adoption of the rifle, he stipulated that the ammunition be made with a different brand of powder, a powder that was evidently manufactured in his domain. Unfortunately, it was a mismatch for the new rifle, and the powder residue clogged up the gas port, causing malfunctions.

That was no problem for the Senator, his pork barrel, or his un-deployed constituents, of course, but it was not a very good thing to be happening for the rifleman in the particular environment in which the rifle was being used. Coupled with the likely circumstances at the time of the malfunction in that venue, it would be liable to cause one's "pucker factor" to "red-line." The scariest thing on earth, as a big-game hunter once observed, is not a *bang* when you least expect it, but a *click*!

We were told about the temperamental sensitivity of the M-16 and shown how to clean the carbon from the little gas tube that protruded back into the receiver, but we heard that some of the Marines up on the DMZ were still using the good old M-1 Garands (cal. .30-06) when they could get hold of them, because of their reliability ("The best battle-implement ever devised," as General George Patton, described it.) The middle of a firefight is not the time to be trying out newfangled gizmos, and a soldier should not be asked to do so. If it is *that* important, let the manufacturer and the lobbyist and the Senator and the Government Purchasing Agent get out there and do the field trials!

Fat chance of that....

At the time of the Vietnam War, battle philosophy had changed in regard to marksmanship. Single, well-placed rounds were not generally considered to be the standard drill as in earlier times. There were night ambushes. There was jungle. There was sweat and fear, and brief, abrupt, violent, up-close-and-personal contact, same as my Father met with in Burma. Hard to see anybody to take deliberate aim at lots of times. What was wanted when contact was made was a large concentration of fire to hold the enemy in place, while artillery and air strikes were called in to destroy him.

The idea of the M-16 was to optimize a weapon system that would deliver the largest volume of effective fire from a package that was still light enough for an Infantryman to "hump the boonies" without wearing him out. Two hundred rounds of .30-06 or 7.62, combined with a nine pound M-1 or M-14, weighs a lot more than two hundred rounds of 5.56 and a five pound M-16. That part was good. But for my part, for the purported advantages aforementioned, I would rather have had the little old M-1 Carbine (cal. .30) instead. They made them by the thousands in WW II. It was simple; it was dependable; and it was battle-tested. These are desirable attributes for a weapon to have. These are desirable attributes for a man to have, also.

The .30 Carbine round didn't have quite the punch as did the 5.56, and it didn't stop the well-padded Red Chinese in winter garb at the Chosin Reservoir, but in the tropics it worked well enough to suit the Marines in the Pacific in World War II, so it should have done fine in Vietnam. In fact it did see service there in the early days with some of the advisers of MACV (Military Assistance Command – Vietnam), and also with some ARVN (Army of the Republic of Vietnam) units. It was as light as the M-16 and a "grunt" could tote as much .30 Carbine ammo as he could 5.56. Most importantly, if the Carbine jammed you could take your muddy jungle boot and stomp on the operating slide handle and get it back in business again. A friend of mine who had served as a "LRRP" (Long Range Reconnaissance Patrol) with the 101st Airborne once expressed his admiration for the little M-1 Carbine because of this trait. In addition, the M-1 Carbine didn't have quite as tight tolerances, I think, as did the M-16, so it could stand up to bad powder and more crud in the works and still function. In spite of the cartridge's lack of authoritative punch, it wasn't much less than the 5.56 NATO round; and it certainly wouldn't make anybody feel *better* if you hit him with it! It was successful in all theaters in World War II, so I reckon it would have done well enough in Vietnam. I'll take a Carbine that goes *bang* over an M-16 that goes *click*, any day. But that is just my humble opinion on the issue. I was not a "grunt." I never had to field-test my theory, but many a poor Infantryman did. They might take issue with me, in which case I will take off my hat to them and bow to their opinion without a quarrel.

We didn't get to pick what we wanted to carry, for the most part. A simplification of logistics is as important as the simplification of anything else when you are operating a war. But later in my tour, instead of carrying my M-16, I sometimes carried a .45 pistol. It was issued in a right-handed holster even though I am left-handed. (The Army does not expect that one should be left-handed while on duty.) Back in the States I fired "expert" with the .45, but it felt like I was shooting a brickbat. The .45 – like the carbine – was reliable and simple, but had been battle-tested for a little longer - since World War I. But I always liked revolvers better than semi-autos. They had those same attributes, but they had been battle-tested since the War for Southern Independence, plus they felt better to my hand. So I would have preferred a revolver in a left-handed holster to go along with my little imaginary M-1 Carbine. I would have preferred a lot of things….

During this week of orientation I lived in a nice little screened-in "hooch". Sharing the hooch with me was a Catholic Chaplain. He had to go through the course along with everyone else, even though he was a non-combatant. Equal opportunity training. But there were no front lines really, much like in the Indian wars of the Old West, and anyone could find himself a target of hostilities, though some obviously were far more exposed to peril than others. A Chaplain had to go where he was needed most, and not all of his "parishioners" were back at the base camp. The good Chaplains often went in harm's way to take care of the men – even, so I'm told, to the point of shielding a wounded man with their own body until a Medic could get to him, and fighting off an enemy soldier with a .45 while doing so. Like the Medics, the Chaplains earned a great respect, and this Chaplain and I became friends. From our brief acquaintance I knew he was a good man and a credit to the Catholic Church, and I knew he would provide good spiritual comfort to the men in his charge.

There was a little library nearby, and I checked out a copy of *Moby Dick.* It was the second time I had read it. My Uncle Dick Neher had introduced me to sailing many years before, and he had given me a copy of Mary Blewett's *Celestial Navigation for Yachtsmen,* which I had just finished reading the preceding week when I was in Nantucket. At night I would read about Ishmael and Captain Ahab while the monsoon rains beat down on the tin roof of my hooch in Southeast Asia, as far away from Nantucket as was the *Pequod.*

After completing the training course, I was assigned a staff job with the 3rd Brigade's Headquarters and Headquarters Company (HHC). My assignment was Base Development Officer, and my duty was to plan and provide for the billeting of the 3rd Brigade units coming in from the field subsequent to their redeployment to CONUS (Continental U.S.). President Nixon had just begun the "Vietnamization" of the war, whereby the Army of the Republic of Vietnam was to take over more and more of the operations as American and other allied forces gradually withdrew from the conflict. The 3rd Brigade of the 82nd Airborne was concluding operations and was to be the next unit out. Men with less than eight months service in Vietnam would not return to the States with their unit, but would be reassigned "in country."

I answered to Maj. Mueller, who had come to the 82nd from the Special Forces – the famous "Green Berets" that President Kennedy had created about eight years before. Their twelve-man A Teams - each man cross-trained in two specialties such as demolitions, weapons, communications, or field medicine – would go out in the "boonies" and train and advise the indigenous guerillas in counterinsurgency operations.

Among those with whom they worked were the Kachin tribesmen of the back country near Laos and Cambodia. I remember my Father speaking of using the Nagas and the Kachins as bearers as well as guerilla fighters in north Burma in their operations against the Japanese. He said they paid the bearers in opium – one cube the weight of an Indian *rupee* per day - and each carried a little set of balance scales to weigh it with. They would pop the opium, test the balance of the pole supporting their load, shoulder it, and away they would go up the jungle trails – presumably with a pretty rosy outlook on things for the rest of the day.

Maj. Mueller had jumped into combat with the Special Forces. He wore the 82nd Airborne patch on his left shoulder, and the Special Forces patch on his right, signifying prior service with that unit in a combat zone. Insignia protocol had recently been changed in the Army to allow for this, and it remains in effect today. Another change made was the subduing of insignia on fatigues. Gone were the white name tabs on the pocket, the white or gold designations of rank, the gold branch insignia, the colorful unit shoulder patches. All were now black on the olive-drab background of the jungle fatigues. All, that is, except for the red, white and blue shoulder patch worn by the "All American, All the Way" 82nd Airborne Division. Others might timidly subdue their unit insignia in hopes of not being seen, but not the 82nd. Let the VC or the NVA see that red patch and tremble!

Speaking of patches, one of the first things I noticed was the number of soldiers without jump wings.

I found out that the 3rd Brigade was not on jump status, but had been reorganized as a Light Infantry Brigade. I drew the jump pay while I served with them because I had just graduated from jump school and thus qualified for the pay for the next three months following, but the unit itself was not on jump status. I don't think the 101st Airborne was on jump status at the time, either, but I might be wrong on that. The 173rd Airborne Brigade was, as was, of course, the Special Forces. I remember seeing a trooper of the 1st Air Cavalry Division sporting an Airborne tab over the Cav's old "horse blanket" patch, and was told that one brigade of the Air Cavalry was also Airborne. So some Airborne units were strictly Airborne, while some others were off jump status and had lots of men who were not paratroopers assigned to them.

I was told - by one who found out the hard way - that at Fort Bragg, an Airborne NCO in the 82nd "outranked" a "leg" 2nd Lieutenant. But that was definitely not the case with anybody in the 3rd Brigade. All of those men slogged through the paddies and jungles together - tasting the copper penny of fear on their tongue, thinking about home and wishing they were there - and fought and died in the mud together. Some lost body parts, lost best friends, lost wives, lost girlfriends, lost faith in their country, and lost faith in their God. But all kept the faith with their buddies, and gave the enemy hell, and brought credit with their blood to the proud old Airborne units in Vietnam. As far as being "Airborne" was concerned, jumping off the skids of a Huey on a hot LZ was as "Airborne" as anybody had to be in the 3rd Brigade!

I had a friend from Forest, Virginia, named Eddy Joe McLane who was a trooper in one of the Regiments. Somehow he found out that I was there and he looked me up. We had a great time, and he would often come around to my hooch for a beer when his unit was on "stand-down" from the field. He would bring his buddy around, too, but I have forgotten his buddy's name. We had a little Officers Club at our compound and one night I gave Eddy Joe and his buddy a couple of my fatigue jackets with officer's insignia so they could go over there with me. It was a small club, and though they got some stares and I got some glares nothing was said. Another time they reciprocated, and dressed me up in an enlisted man's fatigues and took me down to the EM Club. There happened to be a brawl there that night but luckily we did not have to get involved. "Impersonating an Officer," or "Impersonating an Enlisted Man" certainly would have gotten us into enough real trouble to have lasted us for awhile.

Eddy Joe and I used to ride the bulls and the broncs together on the rodeo circuit down in North

Carolina called the Central Rodeo Association. A lot of the cowboys riding in the CRA were from out West and were stationed at Fort Bragg with the Army, or at Camp Lejeune with the Marines. Eddy Joe and I thus had a bond that transcended the little protocols of the Army, and we would get together whenever he was in from the boonies and drink beer and tell each other once again the old tall tales from around the bucking chutes. But one day Eddy Joe showed up by himself. He told me his buddy had been killed on ambush patrol. There was no joy in Mudville after that.

I had started riding in the rodeos to try to impress the girls, and ended up riding in them to try to forget one. My Uncle Ed Luby got me to quit. I was staying with him and my Aunt Betty when I first reported to Fort Belvoir for Engineer Officer Basic. There was a rodeo in Leesburg or somewhere that weekend that I was planning to enter. Uncle Ed said that I was now property of the U.S. Government, and that it was my duty to take good care of government property – and that getting my ass stomped into the mud by a 1500-pound Brahman bull (or "Braymer," as the cowboys say,) was *not* considered taking good care of government property!

Uncle Ed was a thirty-year-career Navy Captain who was at that time serving in his third war – a fighting Irishman as tough as nails, an undefeated boxing champion at the Naval Academy, a sea captain as hard as winter in the North Atlantic, and a veteran of Naval combat aboard Destroyers in World War II. The brand-spanking-new 2nd Lieutenant whose ass was now property of the U. S. Government replied: "Yes, Sir!" – and that was the end of my less-than-promising career as a rodeo cowboy. Uncle Ed told me to "live for something bigger than myself." Good words for a young man to carry with him when he goes off to the wars.

The Phu Loi Airstrip was about an hour's jeep drive north of Saigon. The runway could handle large fixed-wing aircraft as well as provide for the many helicopter units that were based there. It was obviously a consideration in the selection of a base camp for an airborne unit like the 3rd Brigade. The orderly room faced the runway and the Commanding General's quarters were nearby. He was provided with a house trailer, and one of my duties was to see to the construction of a revetment to protect it from the rockets the Viet Cong would send into Phu Loi from time to time.

The orderly room was "mission control." The Commanding General (CG), his Executive Officer (XO), and his staff all had their offices there. The XO had the most soldierly bearing of any man I ever saw. He was tall and straight, broad shouldered and flat in the midriff, had a handsome rugged jaw, iron-gray hair, and carried himself with a confidant bearing and an easy grace. He impressed me as being the prototype of "Airborne, All the Way!" In the building also, I believe, were located the offices of the G-1 (Administration), G-2 (Intelligence), G-3 (Operations), and G-4 (Supply). Whenever you met a superior officer out of doors you rendered him a salute, and instead of saying "'Morning, Sir," you said "Airborne, Sir!" He returned your salute with "All the way!"

The "brass" was always coming and going. I remember one morning the General coming into the orderly room, perhaps after the morning briefing. Behind him came the XO, who bore the rank of full Colonel. After him came two "Light" Colonels. And bringing up the rear of the entourage were a couple of Majors who evidently had not yet established the "pecking order" between themselves, for they both went for the door at the same time and got jammed up for a minute - a little turf war going on back there in the rear ranks.

I shared an office with Maj. Mueller. He was a great guy, and would often defer to my civil engineering "expertise" in our assignments. We had to locate barracks sites, a mess hall, and latrines. We had to make arrangements for water supply, figure electrical generating requirements for lights and air conditioners, and make other necessary provisions for the accommodation of the troops coming in from the field.

Each morning there was a brigade briefing for the General that we had to attend. It was usually pretty crowded, for all the units had to give him a daily Situational Report (SitRep) on their operations. One of the reports that I was always interested in was from Captain Ernie Robke, the Commander of the Brigade's Combat Engineer Company: C Company, 307th Engineer Battalion (Airborne). The Army was interested in giving its Regular officers command time in Vietnam, and I understand that often the drill was to run you through a six-month slot. Thus, if I remember correctly, I think I was to have been next in line for Capt. Robke's job after he went home, had the brigade not been pulled out. It would have been a grand assignment.

The Airborne Engineers had these little D-4 bulldozers and other equipment that was airmobile, if not air-droppable. I know that there is a procedure whereby heavy equipment is secured to skids and put into C-130s for transport. The plane flies low and slow – flaps down - over the DZ, lowers the stern ramp, and the equipment is pulled out with a parachute. It hits the ground skidding. They can also transport equipment slung under cargo helicopters, the same way that the Artillery moves their guns. I don't know how they air-drop bulldozers. They don't bounce like paratroopers.

Anyway, these division Combat Engineer units would send teams out to clear LZ's, string concertina wire and otherwise construct perimeter defenses, destroy enemy bunkers and ordinance caches, bridge streams, cut roads, and otherwise provide engineer support for combat operations. According to the Field Manual for Engineer Operations in Light Infantry and Airborne Brigades, Engineers are considered to be High-Value Targets (HVT) - a somewhat dubious distinction, but one which lends a certain *esprit*.

Engineers have often been considered as reserve Infantry. My Father's brother, Uncle Joe, won a Silver Star when he was killed while serving as Surgeon with the 20th Combat Engineer Regiment in the Hertgen Forest. They had been fighting alongside of the Infantry, blowing up German bunkers. In Burma, Army Engineers were taken off the Ledo Road to make the assault on the Myitkyina Airfield with Merrill's Marauders, and were awarded the Combat Infantryman's Badge (CIB) for their part. I remember my Father saying that "Vinegar Joe" Stilwell didn't believe in giving out medals in the China-Burma-India Theater. He said the CIB was the highest honor that the country could bestow upon a soldier. I agree with General Stilwell. Dad got one of them. Dad has always been my hero, and I have always tried to live up to him.

My assignment was tedious and boring to me, but I tried to do my duty in spite of that. But one day I had a little diversion. I knew the leader of the Phu Loi Patrol, which was a small unit whose mission was to patrol the perimeter for security. On one recent patrol they had found some unexploded ordinance (UXO) in the form of some 155mm artillery rounds. Since the VC liked to find this sort of stuff and make "improvised explosive devices" (IEDs) out of them – as they call those things nowadays - we wanted to dispose of them. The Patrol Leader asked me if I wanted to go out with him and blow them up. I certainly did.

We rounded up the gear, which - besides rifles, ammo, etc. - consisted of a satchel with six two-and-a-half-pound blocks of C-4 plastic explosive, some blasting caps, time fuse, fuse igniters, and a set of crimpers. The two of us then went outside the wire to the various locations where the UXO were located. We cut a test length of fuse and timed it. Then, to rig a charge of C-4 on one of the artillery rounds, we cut the fuse to burn for about five minutes, crimped on a cap, poked a hole in the end of a block of C-4 to stick the cap in, clamped a fuse igniter onto the other end of the fuse, popped it, and went for cover behind a nearby hill. The Field Manual

FM 5-25 tells how far away you are supposed to be for certain charges that you set, with and without overhead cover. When one of the charges went off – which also detonated the artillery round - we discovered the reason for the distance table for overhead cover, after a chunk of iron from the ordinance fell from the sky and buried itself with a resounding plop between the two of us where we sat. We learn as we go.

We disposed of all the UXO and had one block of C-4 left. Walking back along a stream bed we saw a hole in the side of the bank. We rigged the charge to the end of a pole and stuck it down the hole - and if the dreaded VC tunnel dwellers did not thereby learn to fear the might of the U.S. Army Engineers, at least the gophers did.

Behind the orderly room building were our hooches. They were open-air and screened in. In our hooch area there was a small club, or bar, where you could go get a drink or play cards after duty. The main club was behind the hooch area on the road that ran behind the Headquarters compound. There they sometimes had a little band, and there was a bar, of course, and an area of tables. There usually was a good nickle-dime-quarter poker game going on in the evenings that I liked to get in on.

When the units started coming in for stand down, it got right rowdy at the club. There would be

Airborne songs sung and all that sort of swagger, for everybody was going home. A unit would come in, make a commotion for a week or so while the processing was going on, and then they would rotate out and all would be quiet for a little while. Then another unit would arrive.

We were aware of the cultural revolution going on back in the States, but we didn't dwell on that much. The general feeling was that of indifference mixed with contempt. I remember a cartoon that was circulating around just before everyone left. It depicted a little hairy hippie in sandals and bell-bottoms, with a peace symbol around his neck. He was standing by the ramp of a C-130, holding up a sign on a stick which read, "BRING TH' BOYS HOME". Coming down the ramp was a big trooper, with an 82nd Airborne patch on his shoulder, rolling up his sleeve and glaring down at the protester with his fist balled up. The caption read: "Well, I'm home!"

Towards the end of November, 1969, all the units had been processed out. Headquarters and Headquarters Company was, of course, the last to go. Orders were cut and awards were given out. For my modest contribution I was awarded the Army Commendation Medal. I was proud to have earned their recognition, but I am prouder still to be able to say that I once drew jump pay in Vietnam with the 82nd Airborne.

The last night there was a blowout at the club. There was singing and drinking and a general uproar. Out into the night you could hear “Blood Upon the Risers” and “We Gotta’ Get Outta’ This Place” and, as the Holiday Season was close upon us:

> *Jingle bells, mortar shells, Charlie’s in the grass;*
> *You can take your Christmas and etc., etc., etc.*

I remember the next day coming over to the hooch area in the jeep to get the last of my gear to take across the runway to my new assignment. I drove by the Officer’s Club to see what was left of it. Not too much. I remember seeing the door lying out in the middle of the road. Where only yesterday there was abundance of life and bustle, and great events were stirring which would come again no more forever, there was now in the stillness of that morning only empty hooches, and a broken door in the dust left to tell the tale. And so the rollicking, hard-fighting men of the 82nd Airborne all marched away…..

> If I die on the old Drop Zone
> Box me up and send me home!
> Tell my girl I did my best,
> And take me to my final rest!
> Sound Off!
> *One Two!*
> Sound Off!
> *Three Four!*

Take it on down!
One Two Three Four One Two
THREE FOUR!

Patch Crew Working on Road at Phu Loi
Lt. Mike Kinh's 1st Platoon, D Company,
34th Engineer Battalion

CHAPTER III

THE AGE OF AQUARIUS

I arrived in South Vietnam in August of 1969, and Ho Chi Minh died in September. I always figured he must have had a heart attack at the news of my arrival, but there is no way of knowing for sure. He was not a Communist ideologue, but only an Indo-Chinese patriot looking for help from whomever he could get it (2). Ever since the Paris Peace Conference of 1919 he only wanted what Woodrow Wilson, with his "Fourteen Points," had said he stood for – self-determination for all countries. To Uncle Ho, this meant France out of Indo-China - and then us. He'd had a little luck driving out the French at Dien Bien Phu, but by then we had gotten mixed up in the divided situation with our advisers. But in any case, by the end of 1969, Uncle Sam thought that the 82nd Airborne's work there was done, so the 3rd Brigade was sent home soon after Uncle Ho's demise.

Captain Ed Gooch and I were a couple of the leftovers. Ed was from Kentucky, and we hit it off pretty well. Virginia and Kentucky go way back together, from the days when Kentucky used to be one of Virginia's Counties, and from the days of the Virginia and Kentucky Resolutions. Lots of Virginians migrated through the Cumberland Gap to "Old Kaintuck," and I wouldn't doubt that some of Ed's ancestors came from "Old Virginny." After all, there is a Goochland County in Virginia. But anyway, without ever expressing the fact, we sort of felt like a couple of old "homeboys" out there on the far ends of the earth. Just a natural thing – no contrivance required.

Ed told me that there were some R&Rs floating around going to waste and that we ought to take one. R&R is "Rest and Recuperation." Everybody got a one-week R&R during their tour. Thc 3rd Brigade had a bunch allotted that had not been used before the troops went on the "Big R&R" back to the States. Ed said that we could take one "off the record" if we took one during the interval before reassignment. We were not depriving any soldiers (we didn't think about the taxpayers at the time), because, as I said, they were slots that were left over and going to waste – unless somebody had wanted to give up going back to the States so he could go on R&R. Not likely.

Ed's plan sounded good to me. There was only one little problem. One of the stipulations for taking R&R was that you had to have a minimum of four hundred dollars. As I explained earlier, my financial situation was in dire straits. I had been deducting an allotment to the credit card company each payday, so I wasn't drawing much pay. I don't remember how I resolved that problem. I think I had built up enough equity in my credit card by then to take out a second mortgage on it – and also, as a byproduct, to elicit more prayers on the prayer wheel.

One other stipulation was that you had to take plague and cholera shots before you went. The problem with that was they made you sick for a little while, so it was advised that you take them a week or two apart and several weeks ahead of time. I think Ed was squared away on that, but I wasn't, since all this was sprung on me at the last minute. I therefore got both shots together just before we left, which resulted in my being about half sick the whole time on R&R. But Airborne Rangers don't let little things like that interfere with the mission, so I just sucked it up and carried on.

The next thing was to decide where we were going to go. Honolulu was out. That was reserved for married soldiers so that they could meet with their wives. The other choices were Taiwan, Hong Kong, Bangkok, and Sydney. There may have been another, but I can't remember. Ed and I discussed this matter. We felt that if we went to any of the Asian ports we might as well just get in the jeep and go down to Tu Do Street in Saigon.

We had a natural impulse to be with our own kind. Social engineers who would reconstruct a society into one that denies this impulse as a fundamental truth are building their Utopia upon "a damned lie, and nothing other," as Thomas Carlyle said in his Latter-Day Pamphlets; "and like all lies, they are cursed and damned from the beginning" (3).

Ed and I didn't lie to ourselves. To use an insensitive expression prevalent back then, we wanted to go and be with some "round eyes" - so we picked Australia. If we were to express such a reason in public nowadays, the Diversity Police would probably throw us into the Gulag and expose us to a little "sensitivity training" at the hands of some muscle-bound multi-culturalists. But back then (shocking as this may sound) sensitivity training was not at the top of the list on the Army's agenda – probably because back then being a laboratory for Politically Correct social engineering was not its primary mission. Pass the smelling salts, please.

So off we went to Australia. November was Springtime in Sydney, but it was like San Francisco in Winter at that particular time – about fifty degrees and drizzly. At the airport we exchanged our currency and received a briefing on the laws and customs of our host country. We, of course, were guests, and would conduct ourselves as such. I don't think we ever violated the customs of hospitality while we were there.

The Australians are wonderful people. I remember my Father saying that – after stopping over in New Zealand en route to "Injia's sunny clime" (4) in World War II – if he ever had to emigrate from the United States, he would go to New Zealand. I would go to Australia. It is like I imagine the U.S. was in about the year 1910. The people are open, friendly, and totally without guile. In our briefing we were told of some things available to do. One was an invitation to a pub with some of the Sydney folks, which we took. It was a hearty, friendly time.

The hospitality of the Australians knows no bounds, so far as I know. When all the rest of the world is venting against America, Australia remains a loyal friend. They haven't forgotten the close shave the Japanese gave them in World War II, nor have they forgotten the Americans who helped keep them from being overrun. The Australians fought some pretty tough fights in the jungles of New Guinea against the Japanese; and in Vietnam, where they had some troop units deployed in the allied effort, they maintained their long-held reputation for being rugged fighters. The Australians are a tough bunch, and if I had to be in a barroom brawl, I would be glad to have one as a "mate." They don't want you to sit in the corner of the pub and brood in your beer. They want you to get out there in the thick of it and have a jolly old time with the crowd. So we did.

Ed and I got a place near King's Cross, where a great deal of the night life of Sydney was to be found. The first couple of nights we went to some clubs. During the day we went sightseeing in the city. One day we walked down to the Wooloomooloo Docks, where we saw Aborigine longshoremen loading cargo into the ships. The Aborigine are the native Australians, and, like so much of everything about Australia, are different from anybody else on Earth. I had an Anthropology professor tell me one time that the Aborigine are genetically closer to the Caucasians than anybody. They are supposed to have the gift of mental telepathy, and some are reported to be able – with no external means of communication - to tell when a visitor is coming long before he actually arrives. They count their Genesis from the Dreamtime, the time before their insular way of life was altered forever by the evolutions in technology, and by foreign migrations to their land. Now they are given Reservations in the Northern Territories, but some – like these longshoremen – have made adaptations to modern times.

The cold drizzly weather persisted, and as we passed by a little restaurant I saw advertised that they served scones. I had never had a scone, but I had heard of "warm scones and tea," so we stopped in to give them a try. They were a sort of biscuit and, along with the tea, they warmed us up pretty good. I think they might be a sort of Scottish signature dish – like haggis. I never had any haggis before, either, and still haven't. I'm not sure what it is, but I think it is something like boiled sheep's stomach. I have had boiled chitterlings, so I guess I could handle the boiled sheep's stomach, but I much prefer "chittlins" when they are fried up crisp, so I'll pass on the haggis. Another thing that we had one time was a hamburger – made literally from ground ham. Not exactly what I had in mind when I placed the order!

After a few days of the Sydney drizzle, Ed and I started talking about heading up to Brisbane, on the "Gold Coast," to take in some sun, sand, and surf. In Australia, of course, you have to go north to get to the tropics. Brisbane was Australia's answer to Fort Lauderdale. But there was another problem. Regulations forbade us from leaving Sydney while on R&R. Ed and I discussed this and concluded that since we weren't technically supposed to even be on this R&R in the first place, that we therefore technically were not even supposed to be in Sydney in the first place. Thus, we reasoned, since we officially were not supposed to be in Sydney, then - relatively and technically speaking - we would be more in compliance with the regulations if we were not in Sydney. Therefore we felt that a departure from Sydney on our part would be in the better interests of National Policy in the Far East. We wondered how well this interpretation might hold up if presented to someone with more authority on the subject than we had, but luckily no one of that stature happened to be nearby. Thus we were compelled to reach a conclusion on our own; *to wit*: that since the regulations were not clear in this particular situation, the safest thing for us to do would be to interpret them to our advantage.

We started hitch-hiking up the coast road. British Commonwealth countries drive on the left hand side of the road. I have sometimes wondered why that might be. If you think about evolution from the days when knighthood was in flower, you will see that two mounted knights approaching one another would want to meet each with the other on his right, the side of his sword hand - should he have to use it. This custom carried on until the automobile age, long after the reason for its necessity had passed. Just my theory. Anyway, it made for unaccustomed "thumbing," as we used to call it. Not only that, but in Australia you don't "thumb." Sticking up your thumb there is considered an obscene gesture. Luckily we were told of that beforehand. The way you hitch-hike in Australia is to stand facing traffic – which is in the left lane, as I said – and reach across your chest with your right hand and point up the road with your index finger.

On the first ride that we caught, the driver told us that up ahead at the town of Newcastle the road forked. One fork carried you up the coast road, and was frequented by automobiles and local traffic. The other fork carried you up the inland route, which was frequented by truckers, or "lorries." It might take us longer to get to Brisbane on the coast road but there was more traffic; whereas the sparsely traveled inland route might bring us a lorry ride that would carry us straight through to Brisbane. We thought about that for awhile and decided that we would take whichever route the person giving us a lift at the time was taking. Our next ride turned off at Newcastle onto the inland route, and carried us up the back road to the little town of Maitland, in New South Wales.

By the time we walked and hitched about three miles out of town, it was starting to get late in the evening. No trucks had come by, and few cars. We decided that we had better head back to Maitland and see if we could get something to eat and take matters from there. Just after dark a boy about sixteen stopped and gave us a ride back to town. The car was crammed full of boxes of his possessions. He said he had just that day bought the car for five dollars. Ed got into the front and I climbed onto the boxes in back. The young man asked us about ourselves, and ended up taking us out on the town.

The first place he took us was the local pub. This evidently was his hangout, because he knew everybody, young and old. We sat at the bar and, at the suggestion of our host, drank a beer called "Toohey's Old." We ordered it in mid-sized draft glasses, so when the bar tender asked you what you wanted, you asked him for a "middie" of Toohey's Old. We were treating our fine young host. On the bar were baskets of sandwiches, from which we had our supper.

The scene was a lively one. We noticed that there were no women in the pub; evidently the feminists had not made the same inroads in the Outback as in more "progressive" parts. The place was crowded with men, though - and it was animated. There were carryings-on at every table, and there was much interest at the dart board on the wall at the far end of the room where a lively game was in progress.

We were sucking down the middies and taking it all in, when at one of the tables a commotion erupted between two men who each looked to be about sixty-five or seventy years old. Their argument became heated, and their "mates" began to gather about. After about four or five had taken up positions behind each of the antagonists, the two began to roll up sleeves and do all sorts of posturing preliminary to what looked to be impending mortal combat. Of course this was the cue for their mates to step forth and hold them apart from one another and persuade each to settle their differences with an arm wrestling match instead of fisticuffs – much to the relief of the two combatants. So everything came to a halt in the pub while these two went at it with the arm-wrestling – “best two out of three.” Such cheering-on you never heard in your life. All was good sport, and after everything was settled it was proposed that the two clinch their good “mate-ship” by a rousing game at the dart board. This they thought to be a grand idea, and all the mates of one lined up on one side while all the mates of the other lined the opposite. There was cheering from one side as one would make a point, and cheering from the other as their man scored, until all was good fellowship once again and the pub settled back down to its previous dull roar.

Somewhere along the way we found out that a bridge had been washed out down below by a flash flood, and all the trucks were running up the coast highway. This caused Ed and me to rethink our plan to go to Brisbane. We were going to run short of time soon, and backtracking was not going to help. Nor was the prospect of being stuck up the road somewhere and getting back too late to catch our plane back out of Sydney. Our host said he was the ticket agent at the railroad station in Maitland and could get us on the train to Sydney, which came through at 4 a.m. So he took us down to the depot, opened up the office, issued us the tickets, then took us to a dance hall that stayed open until the train came through. I tell you, Maitland, Australia, is a jumping little place.

We made the train and stretched out on the wooden coach seats to try and get a little sleep. Sometime later we were half-awakened by somebody grumbling about these two blokes taking up two whole benches, but nobody boxed our ears so we went back to sleep. About mid-morning we came chuffing into the station at Sydney, and we got off the last train I ever rode that was pulled by a steam locomotive.

The only other thing I remember that we did before leaving Sydney was to go up to a theater in King's Cross and see the play *Hair*. The "Age of Aquarius" had been dawning now for a few years and we thought we ought to go see about it. There we heard all the songs that had become familiar, and heard a sweet young thing sing a sweet duet with some young man about all sorts of creative sexual activities. Then the curtain dropped and rose again and there was the whole cast – men, women, black, white, and everything in-between - standing right out there buck-naked in front of everybody. Of course we had heard about this attraction ahead of time, and we wanted to go and make sure it was true. Then they all sang about the dawning of the Age of Aquarius and exhorted us to let the sun shine in....

Onto the End Times?

The revolution of the '60s was going on full blast, and *Hair* articulated it in a nutshell. It was the dawning of the Age of Aquarius, wherein Peace was to reign with Free Love and Flower Power. Every one would "Make Love, not War," and all would be "Free at last!" from the traditional conventions and restraints and bonds of our society – which evidently included (as Karl Marx said at the conclusion of his *Communist Manifesto*) all other existing social conditions as well – which, boiled down to its essence, of course, is the recipe for Anarchy. Much was said in *Hair* about personal freedom. Not much was said about personal responsibility.

But *Hair* was an eye-opener all right, and it got to me at a time when I was beginning to seriously question some things that, before, I had taken for granted. *Hair* capped the first milestone for me. The Aquarians would have been proud to know that. They probably wouldn't have been as proud to know my considered conclusions, arrived at later.

"Anarchy is the chrysalis state of despotism," said John Randolph of Roanoke (5), arguing with the authority of Classical Antiquity to back him up. The Greek historian Polybius observed that there are three successive forms of government, each with its good and bad aspect, and each giving way to the next in a retrogression: Monarchy brings order out of chaos, but it devolves into Tyranny; Aristocracy deposes Tyranny, but it devolves into Oligarchy; Democracy deposes Oligarchy, but it devolves into Mob Rule and Anarchy, and the cycle begins again (6). A revolution is only a paradigm shift that – if left to run its natural course – leaves the social structure in balance again at its conclusion. But there are those who profit politically from revolution, and who stand to lose when it ends. Therefore, in order to effect an artificial prolongation of the process and preserve their positions of revolutionary power rather than let the revolution run its natural course, they erect a totalitarian government – a Tyranny – over the body politic. This may be accomplished by the State's deliberate destabilization and destruction of civil society (such as church, family, neighborhood, local government, etc.) which atomizes the people and leaves them exclusively dependent on the central government – which is the very foundation of totalitarianism (7). The methods employed to create this dependency vary from fear on the one hand to persuasion on the other - including the promotion and encouragement of a belief in both radical individualism and radical egalitarianism (8).

Such a belief and practice verily defined the radical '60s revolution and the dawning of the Age of Aquarius.

"None but the people can forge their own chains; and to flatter the people and delude them by promises never meant to be performed is the stale but successful practice of the demagogue…." warned John Randolph of Roanoke (9). True Freedom is a noble and precious thing, and is not to be found wallowing around in such as "the flesh-pots of Egypt" that the Children of Israel murmured for during the Exodus. Its rarity carries a high price that must be paid for in the coin of self-denial, not self-indulgence. While those who were paying for it were rolling in the mud in Vietnam and dodging AK 47 fire, the Aquarians were rolling naked in the mud at an indiscriminate groupie-grope called Woodstock, high on drugs and dodging the draft. Cloaking self-indulgence in robes of Freedom, and weak character in robes of self-righteous indignation, they were running away to Canada, burning draft cards, burning flags, burning cities, running for office, and – while better men were gone - getting themselves elected by squandering that which they had done nothing to earn. So much for the freedom of radical individualism.

As for radical egalitarianism, one need only to consider the works of God, "for who can make that straight, which He hath made crooked?" (Ecclesiastes

7:13) God did not crank out mankind on an assembly line, as we are exhorted to believe according to the Communist philosophy and the machine age that spawned it. If "all men are created equal," then Albert Einstein could go fifteen rounds with Mohammad Ali and fight him to a draw, and then Mohammad Ali could sit down with Albert Einstein in the locker room afterwards and intelligently discuss with him the Theory of Relativity. So, when we talk of Freedom and Equality we must define our terms, for the two conditions are mutually exclusive. If equality of opportunity is the goal, that is fine. If equality of outcome is the end promised, then it is a lie promulgated by political demagogues and revolutionaries. God created all men free. He did not create all men equal. It takes a totalitarian government to do that. Let the sun shine in….

When the inevitable time came, Ed and I caught the military charter flight and headed back to Vietnam. We left Sydney in the morning and flew by way of Darwin, in the Northern Territories. I have always liked the idea of Territories. They evoke independence from centralized authority with the very name, and indicate a sparseness of population that appeals to the minimalist in me. In Australia it is only amplified.

Looking out of the airplane window as we flew across the vast Outback, I was impressed by the emptiness. Most of the population of Australia is concentrated on the fringes of the continent. Looking down at the interior I could see nothing for hours but a lumpy, red, eroded moonscape devoid of habitation – except for once every hundred miles or so, a lonely house perched beside a windmill on some remote cattle station way out in the middle of nowhere. How it got there I can't imagine, and what they or the cattle lived on I don't have any idea. Life out there in the Outback was bound to be a struggle for all concerned, but out there is true freedom – as much as one is ever likely to find east of Eden.

At last the plane landed for fuel at Darwin, which is on the north-central coast of Australia. We were told not to stray – the Aborigine on the local Reservation had just run spears through a couple of Park Rangers. Perhaps this was just a story cooked up to discourage desertion – national demoralization by that time was creeping like cancer into the military in Vietnam. But we didn't want to go pay a visit to any spear-chunkers, even if - as the Anthropologist said - they were our next of kin. By then we had begun to be sufficiently enlightened by the Aquarians to eschew at least *those* traditional bonds of society, and go instead to seek a "middie" of Toohey's Old with which to gratify our baser animal appetites. This, with the enthusiasm of the newly-converted, we proceeded forthwith to endeavor.

We were told that the plane was leaving in an hour, and that if we missed it we would have to pay our own way back to Vietnam on the next flight. I would have had to take out a third mortgage on my credit card in that case, so Ed and I checked our watches closely before walking up to the road that led into Darwin. We tried to hitch a ride but had no luck - since there is nothing on the road to Darwin except the airport - so after a little while we decided we had better head back. It was well that we did, for when we got there the plane was already boarding.

We took off, and flew up the Indonesian Archipelago. Down among those lush tropical islands and sparkling waters Joseph Conrad had sailed in his square-riggers and tramp steamers. Looking down at that sea we would not have been surprised to see the rusty old *Patna* steaming westward, carrying Lord Jim to his destiny. But we were bound for our own…..

Personnel Bunker – Fire Base Lobo

CHAPTER IV

FIRE BASE LOBO

Captain E. M. Luby, my sea-captain Uncle, used to say: "You can't force your destiny." But if you can't force your destiny, you can force its resolution. Upon our return to Phu Loi, Ed and I parted company. I don't know what his next assignment was to be, and I don't think he did either at that time, but I never saw him again. I hope he made it back all right to "Old Kaintuck." As for me, rather than wait around for orders to who knows where, I decided to take a hand in the resolution of my own destiny. An old saw says, "Stick to the Devil you know," so I went across the runway and got a job with the 34th Engineer Battalion. I was starting to pay attention.

The 34th Engineers was a construction battalion. Its primary mission was the maintenance and upgrading of the main supply route to the 1st Air Cavalry Division, which at that time was operating north of Saigon. The 1st Cav's mission was to protect Saigon by blocking the infiltration of the NVA along that section of the Cambodian border.

Supplies for the division would come into Saigon by sea and by air, and then would be transported fifty or sixty kilometers (km – or “clicks” in military jargon) northward by truck convoy to the unit’s base camp at Phouc Vinh. Escorted by armor, the convoys ran weapons, ammunition, aviation fuel, spare parts, food, clothing, and all manner of gear needed for the Air Cavalry to conduct its operations. It was necessary to keep this route open and maintained in order that the flow of supplies could continue uninterrupted. This was difficult to do during the rainy season (May to November) when the southwest monsoon turned dirt roads to mud. Highway 13 was paved, but 1A and portions of 2A were not.

The 34th Engineers had been struggling with the problem of keeping the route open during the monsoon rains, but in November, with the coming of the dry season at hand, the battalion was in the process of switching from road maintenance to road construction. Its objective was to improve the worst portions of the route and give it an all-weather surface. This so-called “Highway 1A/2A Upgrade Project” was to be the primary mission of the 34th Engineer Battalion for the next six months.

I was welcomed into the 34th Engineers about the end of November, 1969. To begin with I was given a desk in the office of the S-3 so that I might get acclimated to the unit's operations. The 34th was commanded by a Lieutenant Colonel of Engineers. He was assisted by the Executive Officer (XO), a Major, who was second in command. The battalion staff was comprised of the Adjutant (S-1), the Intelligence Officer (S-2), Operations (S-3), and Supply (S-4). The battalion itself consisted of Headquarters and Headquarters Company, four "line" companies (A, B, C, and D), and the attached 103rd Asphalt Platoon.

The first thing I found out after I got settled was that the Vietnamese barber had blown up the mess hall. The next thing I found out was that the entire ten-man survey party up on the road project had just been attacked and killed by the Viet Cong; and that tractors were hitting mines in the road up there. It sounded like things were a little hairy in the construction business. I wondered what security precautions were being taken.

I remember soon after my arrival being taken around in a jeep to be shown the battalion's various projects - I think by the Battalion XO. On that day I was introduced to the road that would be the focus of my efforts for the next six months. It would be my rite of passage.

The road was across the Saigon River from Phu Loi and to the northeast. At the southern end, Highway 2A began by branching off of Highway 13 at an intersection due east of Phu Loi. From that junction it ran northeastward for about 20 kilometers to where it intersected in a "T" with Highway 1A. That somewhat remote intersection had been dubbed "Claymore Corners" by the convoys that had been ambushed there in the past. (Claymore mines are anti-personnel devices that throw a big load of buckshot. They are stuck up in the ground pointing in the desired direction, and command-detonated electrically.) From Claymore Corners, Highway 1A continued northward for about another 20 clicks to Phouc Vinh. The road ran through old French rubber plantation country and was dotted with villages along the way. Halfway between Claymore Corners and Phouc Vinh the road crossed the Song Be River on a new bridge that had been built by South Vietnamese Army Engineers (Army of the Republic of Vietnam - or ARVN).

The major portion of the construction effort for the coming dry season was to begin with a bad stretch of road on the upper end of 2A, starting about six clicks below Claymore Corners. From there it was to extend up through that intersection and continue north on Highway 1A for another four to six kilometers.

Since Phu Loi was across the river from the road project, to reach the lower end of the project site it was necessary to travel northward on roundabout roads for about twenty-five kilometers. Halfway up this route was Ben Cat, which was on Highway 13 and located on the eastern edge of the "Iron Triangle" – an old Viet Cong stronghold, and the 82nd Airborne's last area of operations (AO). At Ben Cat was the intersection of Route 7B, a dirt road that led eastward about eight kilometers over to the upper portion of Highway 2A. The beginning of our project site was just above that 7B/2A intersection.

The battalion had worked to keep the road open during the monsoons, but at the bad section on 2A where the new project was to begin there had been a great mud hole. It was at a low portion across from the little village of Cao Dinh, where the road ran near a swampy creek. A corduroy section of road had been put in using logs cut from full-grown trees, but the logs eventually sank into the mud, along with a couple of bulldozers that remained buried there until the dry season arrived. In order to keep the route open, a rough by-pass road of about three kilometers in length running parallel to the main road had been cut north of the village and the mud hole.

With the arrival of the dry season, construction work was just beginning on the upgrade, and there was much to do in the six months of dry weather ahead. The first thing needed was a base of operations on the project site. At that time everything was being run out of Phu Loi, and half of each day was being consumed in just getting the equipment up to the road in the morning and back to the base camp at Phu Loi before dark. Next the temporary bypass had to be upgraded to enable it to sustain the convoy traffic while construction on the main road was going on. A borrow pit had to be located and opened to supply the material for the project, and a water point had to be established to fill the tanker that would spray the road to keep down the dust during construction.

Once these preliminaries were completed, work on the project could bcgin in earnest. Culvert construction could begin and the equipment could start hauling dirt for the road. The Battalion Operations (S-3) had calculated the amount of material needed and estimated a Project Schedule to monitor the progress of the construction. Based on this Project Schedule, it looked as though there was enough margin for the battalion to complete its mission in plenty of time before the monsoon rains set in again in April or May. If not, it would be too late, for construction would necessarily be called to a halt by the torrential rains for the following six months.

After a week or so I was given command of D Company. The company was comprised of about 132 men, who made up Headquarters and Headquarters Platoon, the 1st and 2nd Construction Platoons, and a Heavy Equipment Platoon.

Lt. Dennis Casazza, from New Jersey, commanded the Heavy Equipment Platoon. Its Table of Organization and Equipment (TO&E) included the bulldozers, graders, and scrapers – (or "pans," or "earth movers") – and other pieces of heavy road equipment. "Lowboy" semi-trailer trucks were included in the TO&E for hauling the bulldozers.

The two Construction Platoons were commanded by Lt. Jeff Meislin and Lt. Mike Kinh. They had front-end loaders, air compressors, concrete mixers, dump trucks, and a more generalized TO&E of construction equipment.

The Headquarters Platoon was commanded by Lt. John Stienbeck. He took care of administration and supply, and his responsibility was a big one in a construction company. He handled his duties admirably and always with a quiet professionalism. The Maintenance Section was part of the Headquarters Platoon as well - the direct domain of a long and lanky miracle-working Warrant Officer named Wayne King.

The Company First Sergeant was 1st Sgt. Pedro Benavente, who knew everything there was to know about running a company and keeping the Company Commander out of trouble. I called him "Top" – which is short for "Top Kick" - which is what every Company First Sergeant in the U. S. Army is called. But in my book he was also "Top Notch." He deferred to the rank, but at the same time he looked over my shoulder and kept me from going too far astray while he was there. He left before I did, and it was a big loss for me – more than I would realize at the time. Men like him are truly the "backbone of the Army."

As for the CO, I was the "Old Man." I was almost twenty-four years old.

I answered to the Battalion Commander, but S-3 - the Battalion Operations Officer - assigned the projects directly to the companies. Delta Company had three assigned missions at the time.

One was to provide continuing Engineer support to Phu Loi. Lt. Kinh's Construction Platoon remained at Phu Loi to take care of all that. Road maintenance was his main concern, and he had a patch crew working around the base all the time to cover that assignment, and other crews to cover the assortment of other projects that came along. One such project that he had as I remember was the construction of revetments around an aviation fuel dump. All of Mike's work was professionally done.

Another of the company's missions was the Secondary Road upgrade. This was a dirt road connecting the lower end of Route 2A with Route 1A, located about twelve kilometers south of our primary road project. It ran east five kilometers from 2A and intersected 1A at "Dog-Leg Village," a place notorious for being the source of Viet Cong rockets occasionally fired at Phu Loi. A portion of Lt. Meislin's Construction Platoon and some equipment from Lt. Casazza's Heavy Equipment Platoon were assigned to that project. It was a feeder road and was not to be paved, but the surface had to be stabilized, and some culverts had to be put in.

The remainder of D Company – about half of its personnel – was assigned to the 1A/2A upgrade project, which was the company's primary mission. D Company had been given the task of building up with sub-grade the six kilometers of Highway 2A that lay below Claymore Corners, using the locally available soil called laterite for the material. After Delta Company had gotten the operations base established on site, and had started work on its section of the road, C Company was to move up and begin building up the sub-grade on the assigned portion of Highway 1A running north from Claymore Corners. Both companies also had the responsibility of putting in the culverts for drainage at the designated locations on their respective road sections.

When Delta and Charlie Companies had completed the sub-grade and put in the culverts, the road was to be paved. Bravo Company would then put down about a six- or eight-inch layer of rock, or base course, followed by Alpha Company, which would put down the asphalt on top of that. The laterite sub-grade and culverts, the graded-rock base course, and the finished asphalt surface all had to be completed by May, before the monsoon rains set in again. There was time enough, but none to waste.

About this time I heard of another tractor hitting a mine. It was one of the big scraper tractors, with six foot diameter tires, and the mine didn't hurt it – but the operator, thinking maybe he was under attack with Rocket Propelled Grenades (RPG), hopped out of the cab and took off across the field. I didn't know the exact location of all the mining activity, but I was concerned about all dirt roads in general, so on one of my first trips up to the road site with the equipment convoy, I had some men walk ahead up Route 7B with minesweepers. Of course every piece of metal encountered would set the things off and we weren't making much progress in getting up to the job site. I was thinking "better late than never," when an old hand NCO who had "been there, done that" for the past months, tactfully suggested turning a scraper tractor around and backing it up the road while dragging the blade, and letting that mighty piece of equipment detonate any mines that might be encountered. This sounded like a plan.

As we were starting up the road behind the intrepid operator and his tractor, I spied up ahead a long column of Armored Personnel Carriers (APC) heading in our direction in a boiling cloud of dust. When the head of the column came nigh to the head of our column it stopped next to my jeep. In the lead track was the Colonel commanding the column. Someone told me later it was Colonel Patton – son of old "Blood and Guts" himself – but I can't vouch for it, for I didn't ask to see his ID. He took one look at my operation and blistered me with some very creative Anglo Saxon, wondering what in Sam Hill we were all doing tooling along like that behind a funny looking rig running backwards up the road. I tried to explain about mines in the road, etc. He pointed to the Vietnamese locals buzzing past my operation on their mo-peds and lambrettas and asked me in no uncertain terms that if there were any mines in the road up ahead did I think they would be going through my little off-the-wall minesweep operation like that? I saluted and said "No, sir." Then away he went with his Cavalry and his cloud of dust.

I hadn't thought about it that way before, so with that valuable little tidbit of survival information - and duly chastened by a veritable maestro of the art – we turned the tractor back around again and headed on up the road to work. We did, however, put all the big scrapers at the head of the convoy en echelon – just so....

Our first priority was to build a base of operations on our project site. While this was going on, we were to start upgrading the bypass above the village. Both projects required a little land clearing beforehand. The land in that region was as flat as a plate. In times past it had been a rubber plantation, but to help secure the route for convoys by discouraging ambush, the rubber trees had been cleared back about one kilometer from the roadway.

The original clearing had been done by "Rome Plows," which were bulldozers especially modified for land clearing. The blades were set at a 45 degree angle and sharpened along the bottom edge with a grinder. Along the top of the blade there was a bar which was placed slightly ahead of the lower edge of the blade. At the leading end of the blade there was a heavy steel point, or "stinger," that was also sharpened with a grinder. With this blade the dozer operator could stab and splinter a tree, then cut it off at its base with the sharpened blade. The bar at the top of the blade meanwhile shoved the tree away from the bulldozer and kept it from falling back onto the cab. The cab had armor plating around it and overhead - for obvious reasons. The whole rig was contrived in Rome, Georgia, by some equipment manufacturer, and hence the name of Rome Plow.

The plows were operated by independent Army Engineer Land Clearing Companies – and very independent Army Engineers. These men were a colorful assortment of soldiers-of-fortune, some of whom had been in Vietnam for years. Many carried the old M-3 .45 cal. "grease guns" that paratroopers often carried in World War II, because with their collapsible wire stock they were short and handy to keep beside them in the cab while they worked. Although they had a high casualty rate from booby traps and harassment fire from snipers, they suffered a relatively low rate of killed in action (KIA). But most had a Purple Heart or two to their credit, and many had three or four. When they came in out of a month in the woods, they were a wild and wooly bunch. There were several Land Clearing Engineer Companies based around Long Binh or Di An. Legend has it that one company coming in for stand down "treed" the entire 1st Infantry Division. But if these Plow outfits didn't have any Infantry divisions around to pick on they would just go over and beat up each other.

Since the Rome Plows had been up there on our section, a low patchy scrub had grown up which could afford concealment to the Viet Cong. It thus presented a security risk to us, so as we were clearing to widen the bypass road, we also cleared this brush back away from the roadway in the places that we deemed a threat. While some of our dozers were clearing scrub, others were opening a "borrow pit," stripping topsoil so that we could get at material to be used for widening and improving the bypass. Later we would use this pit as our initial source of material for the road project itself.

The material to be used was called laterite, which was a well-graded red soil that was prevalent in the region. "Well-graded" means that the grains of the soil varied in size in a very desirable proportion - from coarse pebbles the size of a pea, downward through a fine grain, and on down into dust. This made it excellent material for road construction, because after compaction there were no interstices left among the individual grains, which made it very stable.

At the beginning of D Company's job site on 2A the road ran northward for about one kilometer, then turned due east and ran five clicks straight to Claymore Corners. We had found some good laterite located on the north side of the road in the bend where 2A turned to the east. It was here that we opened our first borrow pit. After the bulldozers had stripped off the topsoil, the scrapers started loading up with laterite and hauling the material to the bypass, where they dropped it at the designated "lay-down" site, spreading it as they went. A sheep's foot roller, pulled by a dozer, compacted it, followed by the motor graders that cut the final finish on the surface.

The scrapers were Clark 290s, with Cummins diesel engines. They were a rugged and dependable piece of equipment and served us very well. They had four-wheel, rubber tired tractors that could be detached from the pan, or scraper, itself. On the front of the tractor was a hydraulically operated blade, similar to that on the dozers. The blade was used primarily when a detached – or "bob-tailed" - tractor was stationed in the pit as a pusher to help a scraper load itself with laterite. Bulldozers also were used as pushers in the loading pit.

The scrapers had a capacity of seventeen cubic yards of material in their pans. The bottom of the pan opened like a lower jaw and was also hydraulically operated. This jaw was the actual "scraper." To load the pan, the operator began at one end of the pit – at the point designated by the sergeant who was directing the pit operation. When the operator got to the starting point he lowered the jaw of the pan into the laterite, and – while moving ahead - began making a cut, peeling off a layer of material and pulling it into the pan as he went until the cargo hopper was full. If he made the cut too deep it would stop the machine. If he made the cut too shallow he would not have a full load when he got to the other end of the pit. There was a knee on the back of the pan for a dozer to push against and help the scraper to load. With the bulldozer shoving, and the scraper tractor pulling and working from side to side where it was hinged to the pan, the straining of the loading operation reminded one of some prehistoric monster - attended by a midwife - trying to give birth.

After loading, the scraper would haul its load out of the pit to the lay-down site. There another NCO would direct the operator where to open the pan and discharge the material. This he would do without slacking his speed, which would spread the load along a strip of the roadway. When this was completed, he would close the pan and return to the pit for another cargo of laterite. In this fashion, material was put down on a section of road until it was brought up to the elevation indicated by the grade stakes that had been put in by the surveyors.

All the while that the pans were hauling and putting down laterite, the roadway was being compacted by a sheep's foot roller. This was a cylindrical drum about four feet in diameter rolled along the road surface and towed by a bulldozer. Protruding from the surface of the roller and spaced about six inches apart all over were myriad steel studs sticking perpendicular to the surface of the drum. Each was about an inch or two in diameter and about five or six inches long, with a flat "foot" on the end of each. When the roller was passed over soft material these “sheep’s feet” would push down into it and compact it from the bottom upwards. After more passes the roller would begin to “walk” its way up out of it as the material began to compact; until finally the “sheep’s feet” would be walking on the surface of the well-compacted roadway. This was merely a mechanized version of the method used by those consummate road building engineers of antiquity - the Romans – who would actually use a herd of sheep to do the same thing.

After compaction the surface was ready to be graded. Rough grading was always going on during the lay-down and the compaction, but the final grading was where the true artistry was involved. You can teach a man how to run a bulldozer or a scraper – and there are good and bad and indifferent operators to be found just as with anything else – but good grader operators are born, not made. Delta Company was blessed with some of the best in the business.

The bulldozers were the bread-and-butter of the road project. Some rough grading on the road was done with them to spread humps left by the pans at the drop site. They also were used for building and grading the slopes on the sides of the roadway. Bulldozers were used for shoving scrapers in the laterite pit, clearing land, towing rollers, rough-cutting roads, building revetments, spreading dirt, and a hundred other things. Dozers are to a busy road site what tugboats are to a busy harbor. The ones that we had were model D7E Caterpillars, an excellent all-purpose bulldozer that was well suited to our requirements. They had hydraulically operated blades, which – unlike the old fashioned cable blades that had to rely on gravity – could be powered down or up to make a cut, thus giving the operator better control.

While clearing work was still being done on the bypass, a bulldozer hit either a mine or a booby trap. The operator got shrapnel in his back and was evacuated to a medical facility. It was after this incident that we put sandbags in the floorboards of all the vehicles to, it was hoped, give us better protection against mines in the road. Some of us also took to sitting on our flak jackets like some of the chopper pilots did. Whether one wore it or sat on it depended upon one's priorities.

We finally found an effective – if morally debatable - solution to the problem of mines in the bypass road. I did not hesitate to use the idea, but I will admit I was a little surprised at its source. A day or two after the bulldozer had hit the mine, the Battalion Chaplain caught a ride up to the job site with me in the jeep. He was decked out in a helmet and flak jacket, and toting an M-16. I was telling him about the mines that we had been encountering in the bypass road and wondering how the best way would be to deal with the situation. "Simple," he said. "Just run the locals around the bypass. Charlie doesn't want to blow up his relatives. That will put a stop to all that." So we did.

When we got to our road section the next morning, I told Lieutenant Casazza to hold up the convoy and keep all the Vietnamese traffic below the bend in the road until I called back down on the radio and gave him the word. Meanwhile I took a couple of bob-tailed scraper tractors up the road to the bypass junction – which was just west of the village – and had them start backing and filling and making a little commotion just beyond the junction. When all was set up, I called Dennis and told him to come on with the convoy and to let the local traffic through.

A few minutes later about thirty or forty people on their mo-peds – men, women, and children - came up to where I had the tractors stirring around, and we flagged them off the main road onto the bypass. As I saw them go, I distinctly remember hoping that none of them would be hurt. I am glad to say that no one was. But I was also fervently hoping that this would put a stop to the mining, and prevent any more of our men from getting hurt. Evidently the word got out to the Viet Cong, for – with one notable exception a little later on – we didn't have any more problems with "Charlie" putting mines in the bypass road, thanks to the idea given me by our "Fightin'

Parson." But I'm the one who made war on innocent civilians that morning, and I reckon I'll have to answer for it to Somebody one of these days.

While we were opening up the bypass road, we also were clearing off a location for our base of operations. This was to be situated about halfway down the bypass road and on the north side of it, across from the village. After the bulldozers had cleared off the scrub brush, the scrapers started hauling in dirt to build a circular berm, which would mark the perimeter of the proposed base. The base would be maybe a couple of hundred yards in diameter, with the berm around it being about fifteen feet wide and about six feet high. This was to be the rampart of our fort, or Night Defensive Position (NDP), or Fire Support Base (FSB), or - as we were to call it - our "Fire Base."

We used the topsoil in the area – a white powdery, gritty sort of dirt - as the material for the berm. The scrapers would climb the ramp on one end, run around the top of the parapet while discharging their cargo, then run down the ramp on the other end. When the desired elevation had been reached, bulldozers dressed up the ends and cut a gate, or sally port, through it, facing the bypass road.

As the elevation increased the width on top of the parapet narrowed, until handling those lumbering scrapers up there called for some pretty close work. On one of the last passes, when the berm had reached the height of about six feet, one of the scrapers got a little too close to the edge. It slid down the slope sideways and rolled over onto its side, dumping its load of dirt. The operator was unhurt, but the cab on the tractor was mashed over to one side. Lt. Casazza got some men together and rigged winch cables from a couple of dozers and pulled the monster back upright. Wayne King and his maintenance crew had it hauling dirt again in three hours' time – with its undaunted operator back in the saddle.

After the ramparts of our fort were completed, we began stringing the perimeter wire and constructing bunkers. Lt. Meislin's Construction Platoon was primarily involved with these fire base tasks while Lt. Casazza's Equipment Platoon was building the bypass road. On the outside of the fort, coils of concertina wire were strung and pinned in place with engineer stakes. Concertina wire comes in coils about four feet in diameter. It is pulled apart, much in the fashion of the musical instrument of that name, and the coils are stretched in rows around the perimeter to prevent infiltration by the enemy. The wire itself is like a continuous flat blade about a half-inch wide with razor blade protrusions along the edges. We set several rows of wire around the perimeter and away out from the parapets of the fort. Across the entrance to the fire base –

which faced south onto the bypass road – the wire could be detached from the engineer stakes and contracted to let equipment and personnel through, and stretched out again at night to close off the perimeter.

On the parapets we built three bunkers, protected with sandbags and overhead cover. They had an entranceway by the inside of the parapet, and a narrow opening above it facing outwards for observation and defensive fire. These were to be manned at night for security. We built one on the southeast side to cover the gate. Another we built on the northeast side to cover the open area to the east and the far line of rubber trees to the north. The third bunker faced westward, covering the long stretch of flat plain that lay between the rubber trees and the road.

After these guard bunkers had been built, we began work on the personnel bunkers for the troops who would be operating out of the fire base. The Infantryman is issued a shovel with which to construct himself a hole in the ground, and digging is one of his many accomplishments. But the lucky Engineer has a bulldozer to dig his holes for him.

We constructed four-man sleeping bunkers as follows: first we took a dozer and scooped out a trench the width of the blade, the depth of a 55 gallon drum standing on end, and the length of two cots placed head-to-toe. Then we stabilized the sides and ends of the trench using sheets of airfield landing mat held in place with engineer stakes driven into the ground. Meanwhile, Jeff Meislin sent his dump trucks down to the 103rd Asphalt Platoon at Phu Loi and started hauling as many empty asphalt drums as he could get his hands on back up to the fire base. These we stood in a solid row down the centerline of the trench. The tops had previously been cut out of them for removal of the asphalt, which made it easy for us to take a front-end loader and fill them with dirt after they were in place. This divided the 4-man bunker down the centerline and left enough space on either side of the row of drums for two cots to be placed end-to-end.

For constructing the overhead cover we began by using corrugated culvert halves placed in a row to make an arched "roof" over each side of the bunker. One end of the culvert halves rested on the asphalt drums running down the centerline of the bunker, and the other end rested on the ground at the side of the bunker. At each end of the archway of culvert halves a sandbag revetment was built across the top, while landing mat set on edge was staked down the sides to contain the dirt. Then, to complete the cover and make it "bomb-proof," a bucket loader piled dirt on top of the whole concern.

The individuals who were to occupy the bunker would then dress up their entranceways. After cutting a couple of steps down into the ends of the trench for access, they would square things off and build a sandbag revetment across each end to provide cover for the two openings to the bunker. When this had been accomplished, the occupants could then enjoy some of the finest bunkers to be had in all of Southeast Asia.

The bunkers were located in rows around the perimeter of the fire base inside of the parapet. This left an open area in the center of the fort for the equipment. To the north of the equipment park was the maintenance tent, which was large enough to accommodate a scraper tractor. To the north of that was the command and communications bunker, where we kept the radios and the assorted administrative items needed for operations. Behind that – and along the north wall - we had the latrines, the water tank, the shower, and the mess tent and chow area. The medic tent was to the south, near the gate.

With all this completed, we moved in about the middle of December. It was a very fine fire base – reminding one of an old British hill fort of Celtic antiquity. When we had put the final touches on it, we hung out our shingle for business:

Welcome to

FIRE BASE LOBO

Home of

"Dusty Delta's" ROAD GANG

D Company, 34th Engineer Battalion

Lots for sale

With the exception of the ARVN guard at the Song Be Bridge, we were the only friendly outpost on the forty kilometers of road between Phu Loi and Phouc Vinh – a glimmer of light on a darkling plain.

The Road Gang at the Secondary Road

CHAPTER V

THE ROAD GANG

Why Charlie didn't come in there and booby-trap the place when he had the chance, I'll never know. It didn't take us long to build Fire Base Lobo, but during that time we were still operating out of Phu Loi. Charlie could have easily slipped in there at night and fixed us up pretty good. But Communists have to go to a lot of meetings, so I suppose they were busy with all that instead.

With the Vietnamization process going on, I was told that South Vietnamese Army troops – the ARVN - were operating in our area. I never saw any of them, nor did I know anything about their operations. We never were in contact with them at all about anything. One day while coming up from Phu Loi I saw the bodies of three dead Viet Cong beside the road, their hands tied together with commo wire. Another time I was told that an ARVN patrol had intercepted some VC heading towards our borrow pit. That of course was a good thing for us, but except for the guard posted on the Song Be Bridge, that was all the evidence of their presence in the area that I was ever aware of.

There were still some American units operating in our area. They would show up at the fire base from time to time and stay with us for a week or two before moving on. Once an Artillery Battery came in for a brief time, fired their guns, and then went away. Elements of the 1st Infantry stayed with us on several occasions, and once a Squadron of Armored Cavalry operated out of the fire base for a week or so. And once my Brother Rat Bob Green passed by with some of his men in the 1st Combat Engineers. I told him I thought that was a hell of a way to make a living, poking around in the dirt for land mines. But for the most part – until C Company came up to the fire base sometime in January - we were on our own.

The first thing we did was to establish security. We set up a guard roster for securing the fire base and the job sites. Guard duty was for three days at a time. At night guards were posted in the three bunkers on the parapets, with two men in each. I would have the Sergeant of the Guard come and wake me sometime during the night so I could go around the fire base and check on things, and see that everybody was alert in the bunkers.

During the day we posted a guard at each portion of the project. The Company had in it's TO&E three

M-60 machine guns. We posted one at the borrow pit, one at the water point, and one at either the lay-down site or at the culvert site – whichever was the more vulnerable at the time, depending on its location. Each guard also had with him a portable battery-powered radio to keep in communication with me, with the Platoon Leaders, and with the NCOs. I had an AN/VRC-47 radio that was mounted in the jeep, and the Platoon Leaders had radios in their three-quarter ton trucks. We did not have any machine guns down on the Secondary Road project, but there were radios, and all the men were armed, of course, as they all were on the primary project site. As those on the guard roster would be excused from their construction duties, and as it was imperative that we keep all the heavy equipment operating on the road, the larger portion of the guard duty fell upon Lieutenant Jeff Meislin's Construction Platoon.

We had a Medic at the fire base, and during the day he would be posted out on the job site with his three-quarter ton truck, ready whenever needed. All soldiers love the Medics, who are all invariably named "Doc."

After these immediate security measures had been put in place, I contacted an Artillery outfit at Phouc Vinh, got their radio frequency, and set up some pre-planned fire co-ordinates for the vicinity of the Primary and

Secondary Road projects, and for the fire base. I hoped that things would not get to the point where we would have to call in any Artillery on Charlie, but it is "better to have it and not need it, than to need it and not have it." Pre-planned fire co-ordinates let the Artillery figure in ranges and trajectories ahead of time, and give the unit that calls it in points of reference from which to adjust fires onto desired targets. This simplifies and expedites things at a time when simplicity and expeditiousness are usually at a premium.

Our greatest vulnerability – and one which we could do little to change – was our pattern of operation. In deer hunting, one seeks to discover a predictable pattern in the behavior of a buck by locating his bedding and feeding areas and finding his route of travel between the two. Once this has been established, the hunter then tries to ambush him along his travel route in the evening when he comes out to feed, or at dawn when he returns to his bedding area. Our pattern was opposite from that of a deer: we would "feed" during the day on the road, and "bed" during the night at the fire base. Granted the Road Gang was dangerous quarry, but Charlie was still the hunter; and we were in his home territory. The U. S. Army had been there for four years and claimed control of the area. Charlie had been there since the beginning of time – and he wasn't going anywhere.

I felt we had done the best that we could. We had security posted day and night. We kept our arms at hand and our powder dry. We had good communication. We had artillery we could call on if it got that bad. We had Medics we could call on if it got even worse. But we also had a job to do, and everybody knew that it was what we were there for, in spite of the dangers. So coming out of Fire Base Lobo in the mornings we would let the big scrapers run mine interference for the rest of us, and then we would go to work. The 1st Air Cavalry Division needed the supply route, and we had to get it built for them before the next monsoon.

When we first began working out of the fire base everything was pretty much concentrated on the lower end of the road section. We had our first culvert site down there next to where the road equipment was working at the lay-down site. Our laterite pit was at the bend in the road about a kilometer above the lay-down site. Behind the pit ran a creek, where we established our water point. This was non-potable water, of course. It was used only for filling the tanker truck that sprayed the road to help keep down the dust. This water point served us throughout the project.

In the beginning, the road work progressed towards the pit. The closer we approached, the less time it took for the scrapers to make a circuit between the two locations. Furthermore, the land was relatively high on that section, and the road there did not require as much material for the sub-grade as it would on the lower portions farther on. With the low volume of material required on this section - combined with the quick turn-around time of the scrapers - work progressed rapidly, and by Christmas we were up to and around the bend, having completed almost two kilometers of the project.

Christmas at the fire base was just another work day. I checked on that with the Battalion Commander beforehand. But I "improved on the orders" a little bit and interpreted "work day" to include equipment maintenance Christmas morning. The Road Gang pulled maintenance in the morning, and then went out on the road in the afternoon and hauled dirt. I thought both the maintenance and the respite paid dividends, but that was not the issue. The Colonel called me into his office and explained it to me sort of the way "Stonewall" Jackson once explained it to General A.P. Hill. Hill's reason for changing the order of march was excellent, Jackson said, but he should have obeyed the order first and reasoned about it afterwards. Then he put General Hill under arrest.

I wasn't put under arrest, but I was given the first in a subsequent series of little "adjustments." I had acquired the bad habit of trying to split hairs with the regulations from my days as a Cadet Private, and it had earned me plenty of trouble at VMI. It would soon get me into plenty of trouble with the 34th Engineers as well….

Everybody had thoughts of home. As we were half a world away, and ahead on the calendar, we were spending Christmas Morning while our families back home were spending Christmas Eve. Mom would be making mince pies for dinner the next day, Dad would be bringing firewood onto the back porch, my brother would be out of high school, and my sister would be home from Hollins College. My old coon hound would be asleep in front of the fireplace – as fat as a butterball, and spoiled rotten by my Mother - while down at the stables the horses would have been fed and put up snug in their stalls for the night. They say at midnight on Christmas Eve the animals talk. Maybe they tell about the time when their ever-so-Great Grandpa stood by a manger in a stable in Bethlehem. Out in front the pasture would have turned brown; and looking down the hill across the creek towards the woods and fields where I used to roam, the evening would be settling over the far ramparts of the Blue Ridge Mountains stretching away to the southwest against the dark Winter sky.

One of the men at the fire base came up to me with an envelope. It was a letter from his Mother and addressed to me. In it she told me what a fine young man her son was, and asked me to see that he would be all right. Of course I answered it with care. I told her that her son was indeed a fine young man, and a good soldier. I told her that he was serving with good soldiers – men who would do their Duty and not let each other down. I told her that we were caught up in a great tide of historical events that were beyond our control, but that we were doing the best that we could to perform our mission credibly and with Honor. I told her that every man there had a Mother that felt as she did, and that I was trying my best to do everything in my power to see to it that everyone was taken care of. But I had to tell her also that if I gave special consideration towards one, it would have to be at the expense of jeopardizing another, and that I had to do the best I could for all in an equitable fashion. At best, I know, it gave only cold comfort to a Mother helpless to do anything for her son. But as it turned out, it was the only letter I ever had to write home to a soldier's Mother, and I am so very thankful…..

The good Chaplain did not neglect us. He would come up to the fire base periodically and conduct services in the maintenance tent. The first time he came up I rounded up the whole Road Gang and herded them into the tent. I figured we needed all the help we could get, so there we all were – Methodist, Baptist, Catholic, Jew, Gentile, Infidel, Atheist, whatever – all singing Protestant hymns. The next time the Chaplain came up some of the troops reminded me about the separation of Church and State and, of course, I did not press the issue. But we had made a good friend of the Chaplain, and he, in turn, put in a good word for all of us where it must have done some good.

Soon afterwards, while I was down on the Secondary Road project site, I got a call on the radio that Lt. Casazza, the Equipment Platoon Leader, had hit a mine on the bypass road. Charlie had placed it in a pot-hole that the tractors had missed. There is no telling how long it had been there lying in wait for one of us. Fortunately, nobody was killed. The Medic was taking care of them, and the Med-Evac chopper was on the way.

By the time I got up there the chopper had landed and they had put the driver aboard. Dennis was able to walk, but his back was injured. We got him aboard and the helicopter took off. Luckily their injuries were not as serious as they could have been, although the mine had blown the front end completely off of their three-quarter ton truck. The sand bags in the floor board had saved them. Details, details, details – and don't get complacent. Dennis was back on the job in a few days, and the driver soon thereafter – both having been awarded the Purple Heart.

The Purple Hearts that the government has in inventory, by the way, were all minted in 1945 near the end of World War II, preparatory to the invasion of the Japanese mainland. It was anticipated that a million would be required, but the atomic bombs precluded their issuance. My Uncle Dick was on the ship bound for that invasion, but he was diverted to the Philippines when the bombs were dropped. Chances are those terrible bombs saved his life - and that of my Father as well; for he would have been re-deployed for that invasion also. In that event the patient readers of this tedious deposition would most likely have been spared this harangue. But in a vault somewhere there are a million Purple Hearts bearing mute testimony to what might have been.

In January was the Lunar New Year, celebrated in Vietnam as the Tet Holidays. During Tet of 1968, the North Vietnamese Army (NVA) launched a major offensive in South Vietnam. My cousin, Jim Traywick, was serving as the Commander of an Ordinance Platoon at the time, and he said things heated up considerably for awhile. Tet of '69 was not nearly so bad, but we wanted to take extra precautions when Tet of '70 rolled around anyway. We had the men clean their weapons each evening when they came in off of the road. We had them wear their flak jackets and helmets when they stood in the chow line at the mess tent, and had them space themselves six feet apart so if a mortar round landed in their midst it would minimize the number of casualties. We had to keep pestering them about that, but they did it.

One night Charlie fired about a half-dozen mortar rounds at the fire base, but they landed in the perimeter wire and not inside the fort. I ran to the Command bunker and found Lieutenant Casazza already there. I didn't know if Charlie was fixing to make an assault or not. We had a "starlight scope" out there in the northeast bunker. It was mounted on an M-16 that we kept with the guard on the parapet. It gathered and amplified ambient light, so that one could literally see in the dark. It gave a green picture of everything. Dennis said he would go check it out, but I told him that I would go, and for him to stand by the radios. I scanned the bush outside the perimeter for awhile, until I concluded that the VC had just been taking pot shots at us from the rubber trees. For all of our concern, that was about the extent of the excitement for us around Tet of 1970.

After a hot day on the road, the men would come in and check the oil in their equipment, grease it, blow out the air filters, and do whatever else needed to be done to have it ready for the next day. They could clean themselves of the dust at the shower point, and then get a hot meal from the mess tent. The men worked seven days a week from "can't to can't" (can't see in the morning to can't see at night), and during the whole dry season they only got Christmas morning off, and one other morning off when a helicopter was sent up to the fire base to let the men take a ride and see things from the air. I thought that a grim life like that needed a little spark to give them something to look forward to, so we set up a beer cooler and allowed a ration of three beers per man per night, just to cut the dust. It was sort of like the grog ration in the British Navy. Every day the Mess Sergeant or one of the NCOs from the Platoons would make a run to Phouc Vinh in a three-quarter ton truck to get supplies. We had a "kitty" for the beer – the men would pay for it at cost - and the supply would be replenished at the PX from that fund each day. It worked just fine and nobody abused the system – unless you count that time when the monkey got drunk.

Soon after Tet, a Headquarters element from the 1st Infantry Division came in to the fire base and cut out all the beer for the time they were there. We had to share our bunkers with them, also. It got a little crowded and it got a little thirsty, but they repaid us for our hospitality by giving us a show every evening. They had this little performance that they called "The Mad Minute." This was an exhibition of firepower like I had never seen before. Sometime in the evening or during the night – it was randomly done – they would line their guns along the north parapet and cut loose with everything they had at the line of rubber trees. The purpose of this exercise was to discourage the Viet Cong from getting any big ideas back there in the woods. It is hard to vouch for the effectiveness of preventive measures sometimes, but if you've got it, use it.

The 1st Infantry had these track-mounted sawed-off double-barreled 40 millimeter anti-aircraft guns that fired high explosive rounds. They were cranked down to zero elevation, and used for direct fire onto the target. It seems to me that they were fed ammo from something like stripper-clips, similar to how the old Mausers and Springfields used to be loaded. They would keep up an awesome rate of fire for about a minute, throwing anti-aircraft explosives in a flat trajectory and tearing up the rubber trees. This drill was well-designed to ruin Charlie's whole day if he got caught in the wrong place at the wrong time. This was the Mad Minute, courtesy of the old 1st Infantry Division, the "Big Red One," (motto: "If you're gonna be one, be a Big Red One!") who cared to send only the very best.

But one night, not long after the smoke from the Mad Minute had cleared away, I heard from the tree-line the defiant "chuff! chuff!" of two mortar rounds coming our way. Some scrawny, half-starved little Charlie had humped those two mortar rounds all the way down the Ho Chi Minh Trail from North Vietnam, plodding along for weeks in his flip-flop sandals that he had made out of some worn out truck tires, just so he could lob them at us. I thought then that it was going to be a pretty tough row to hoe to whip somebody like that – Mad Minute or not. And after all, we didn't.

Our intrepid Chaplain came up there to be with us one day during this time, but somebody forgot to tell him about the Mad Minute. But not to fear; that evening, when things cut loose like Armageddon, our "Fightin' Parson," - like all good men of the cloth - was ready for Judgement Day. Out from his bunker he charged in full battle array. Armed with the Gospel of his M-16, and shouting his war cry of "Incoming! Incoming!" he made for the ramparts like Joshua going up from Gilgal to smite the Amorites.

Our Chaplain was an interesting man, and utterly fearless. I wish I had taken more time to get to know him better. He was not only Airborne-qualified, but a Master Parachutist. I don't know how long he had been in the service, but he was older than most of us. I don't know if he had been in the Chaplain's Corps all that time, either, nor how long he had been commissioned – for he only held the rank of Captain. He had an old-fashioned view of things; he once comforted one of our men whose wife had deserted him by telling him: "Son, why don't you re-enlist?" He must have been an old Romantic who felt as the Victorians did – that the best way to forget a woman was to go off and join the French Foreign Legion. I had the feeling somehow that he was a lonely man himself.

Perhaps something like that had happened to him. But he could preach a good sermon, straight from the Gospels – and straight from his heart. He was a good man.

After we had been at Fire Base Lobo for about a month, C Company came up and joined us. I had known the Company Commander from jump school at Fort Benning. To accommodate another company, the fire base had to be expanded. Charlie Company built a separate area and put a berm around it as we had done. It was attached to the eastern side of our existing parapet. It doubled the size of the fort, and made it like Siamese Twins – two entities joined at the hip. The partition was left standing between our two areas, and C Company cut their own gateway into their area. This separation was not because we didn't like each other – we walked back and forth over the berm to visit all the time – it was just too much trouble to tear it down. Besides, if the VC were to ever get inside of one area, they would still have a fight on their hands trying to get into the other.

As soon as Charlie Company got settled in, they began work on their section of road, which, as stated earlier, began at Claymore Corners and ran north up Highway 1A. Meanwhile we were progressing eastward towards the point where the bypass turned off from 2A, just west of the village of Cao Dinh. We had passed our laterite pit and every scraper load thereafter carried us that much farther away from it. In addition, we were beginning to get to the bad portion of the project where the mud hole had been that had given so much trouble during the monsoon. Although it had now dried out, this low portion would require a great deal more material. We were still hauling plenty of dirt, but linear progress along the road was necessarily slowed the farther along that we got.

To speed up the turn-around time of the scrapers, we searched along the road for a new source of laterite that would be closer to the lay-down site. We found some good material to the south of the road just to the west of the bypass junction, and we opened a new pit there. It turned out to be an excellcnt source for us, and it supplied us with laterite for the rest of the project. The last mine the Road Gang hit was in the road that turned into the new pit. Charlie liked newly turned earth, for it made it easier for him to dig – and harder for us to tell that he had been digging. It was detonated one morning by one of our dauntless scrapers. No one was hurt.

I tried to learn how to operate some of the equipment, and took a hand at the dozers and scrapers. The scrapers gave a pretty wild ride in the pit. With their struggle and their enormous tires they would sometimes get to bouncing pretty violently. The seats were cushioned with springs, and sometimes harmonics would combine and try to toss the operator out of his perch. I don't remember if they had seat belts, but those scraper operators were some pretty good cowboys and could hang on till the whistle. This was a good thing, for an OSHA inspector would have probably had a heart attack on Highway 2A long before he had gotten around to inspecting for seat belts anyway.

One day Sergeant Stiggers, a tough road foreman who sort of tolerated young officers – including Company Commanders - but didn't tolerate any foolishness from anybody, was up there in the "shotgun" seat of a scraper tractor delivering an harangue about something to the PFC operating the rig. They were in the pit loading, and the tractor started bouncing. It catapulted Sergeant Stiggers up out of the seat in mid-harangue, ramming his head against the roof of the cab, which in turn knocked his false teeth out of his mouth. They hit the deck and bounced out of the door. Before Sergeant Stiggers could do anything, the PFC getting the lecture scooped them up into the cargo hold of the pan with the load of dirt. For days after you could find Sergeant Stiggers poking around the lay-down site, looking for his false teeth. They are still over there, buried somewhere in the sub-grade of Highway 2 Alpha, about four clicks west of Claymore Corners.

Sergeant Stiggers was but one of the Road Gang's excellent Non-Commissioned Officers. I remember Sergeant Baker, Sergeant Cruz, and Sergeant Mendoza – good ol' Sergeant Mendoza – very well. They were the ramrods of that road, and it never could have gotten built without the likes of them. The Army can ill-afford to lose men like that, but it was losing one: Sergeant Cruz told me he was getting out when he got back home. He had eighteen years, and was going to give up his retirement prospect by not staying in for twenty. Too much politics. I couldn't blame him. I was getting out too.

Behind the scenes of the big scrapers and their dirt-hauling – and of all the other equipment operation – was Chief King's Maintenance Section. Wayne kept everything rolling, and without him and his crew, all the engineering in the world would not have put down a shovel-full's-worth of dirt on the road. "All you need is fuel, air, and compression to keep a diesel running," he once told me, and I have never forgotten. He supported us from his big shop at the company base at Phu Loi, but he sent two of the best mechanics I ever saw up to the fire base.

I very well remember Junior as one, although I don't remember his last name. He was from Tennessee, and when he was in a talkative mood he might say five or six words. He was one of those reticent mountaineers whose great-great-grandpa had fought at King's Mountain, or maybe gotten tired of the crowds up there in the hills and went off to Texas with Davy Crockett to fight at the Alamo. We had another Tennessean with us named "Tiny," but I can't remember if he was the other mechanic or not. Tiny, of course, was the biggest man in the company – if not the battalion.

Initially the mechanics were under the charge of an E-6, but he required a little too much lubrication himself to do us any good up there, so he was sent back to Phu Loi where Wayne could keep an eye on him. The two Spec/5s that had come up there with him ended up carrying the load all by themselves. And carry it they did. Many nights they took a piece of equipment into the maintenance tent and did not stop work on it until it was ready to go – with the sky coming light in the east. Up there in the grit and dirt and the poor light, and often with make-do tools and patched-up parts, and home-made things fabricated with a piece of scrap metal cut with a torch and welded up to make something work, those two men performed mechanical miracles – and they kept the equipment rolling. I really don't know how they did it.

Putting down dirt was more dramatic, but not more important than putting in culverts. Water is the enemy of a road. Graders cut a crown on a road so rainwater will run off of the surface. Culverts let water on one side of a road go to the other whenever it wants to. The water is going to go there anyway, and if you don't give it a way that suits the road, it will go to suit itself – in which case it will take the road with it when it goes. Six clicks of road aren't worth anything if a culvert washes out and cuts it in two. It has the same effect as blowing up a bridge.

North of the road was the creek – the Suoi Bong Trang – on which was the water point. Its source was behind the rubber trees to the north of the fire base, and it flowed southwestward towards the bend in our road, then turned westward on its long way to the Saigon River. To the south of our road section was a swamp, which was the source of a creek called Suoi Cau Dinh – which I suppose gave the village its name. This swamp was below the middle of the east-west portion of our road site, and south of the village. Its proximity was the source of the monsoon mud hole.

Drainage across the road was generally from north to south, and we had to establish some culvert points along the road to allow water to drain across without destabilizing the road bed. If I remember correctly, we had three culverts on our section. The first, as mentioned before, was on the lower end, south of the bend on our portion of the road project. The second was between the bend and the intersection where the bypass departed from

2A. The last was at the big mud hole east of the bypass junction, and on the western end of the village.

Lieutenant Meislin and his Platoon were the culvert constructors. To install one of these culverts, it was first necessary to divert traffic around the site. This was done by building a temporary bypass around it, and over a temporary culvert pipe. After a bypass had been established, a cut would be made across the main road deep enough to allow for the foundation of the culvert to be built. This foundation was built of compacted laterite, set to a certain slope and elevation to provide for drainage. The culvert pipe would then be assembled out of halves bolted together, and rolled into place on the foundation.

A dump truck and a front-end loader were on site to keep the culvert site supplied with the laterite needed for "backfilling" around the pipe, but plain old shovels were primarily employed in the actual placement of the backfill material. As the backfilling was going on, the material would have to be compacted as it was built up around and over the culvert. This was done using a hand-held pneumatic tamper called a "pogo stick," which was a reciprocating piston with a club-foot on the end of the piston rod that thumped the ground when the operator held the lever down. An air compressor provided the power source. All of this was labor-intensive and time-consuming, and as time began to press upon us, Vietnamese locals were hired for much of the labor involved in the compaction process. A culvert site in full swing might have a half-dozen *papa-sans* out there doing the "one-legged two-step" with their pogo sticks, dancing to the music of an air compressor all day long.

While backfilling was going on, concrete abutments on each end of the pipe had to be constructed to hold the sub-grade of the road in place where it passed over the culvert. Carpenters would first construct forms for the concrete, and fabricate the "re-bar" iron that would be incorporated within the concrete for re-enforcement. When all was braced up, the concrete would then be mixed on-site and poured into the forms. After the concrete cured, the forms had to be stripped back off. The curing of the concrete took time, and of course no backfilling could be done at the abutments until after the forms had been stripped.

Ideally, the culverts should be completed ahead of the lay-down site. This would allow the material that was being put down for the sub-grade to cross the culvert without interruption in the lay-down process, aiding in its uniformity of compaction. It would also keep the scrapers from having to break stride to go around the culvert bypass, then pick their speed back up again. But with the security demands that had been placed on the Construction Platoon, combined with the time-consuming processes involved in construction, we were always behind schedule on the culverts. But I felt that an incomplete culvert site would be a little more forgiving than the uncompleted road itself if we got caught by the monsoon. We made do the best we could with that aspect of the project, but everything we had to do was involved with a question of priorities. However, at that point, it looked as though we would be finished with our road section in plenty of time – unless some unforeseen circumstance intervened to prevent it.

One day a Huey helicopter landed on our road site near Lieutenant Meislin's culvert site. A two-star General - arriving unannounced - disembarked and walked over to me. I saluted and gave him my name. He introduced himself as General Dillard, Commander of Engineer Troops in Vietnam, and said he was inspecting some of the Engineer projects in the region. He asked me some things about our project, and then chatted a little bit with me, asking me where I had gone to school. I told him VMI, Class of '67. He said he was VMI Class of '42, and asked me how things were at the Institute the last time I was there. I told him what I knew. "Ah, well," he sighed - as all older VMI men always say to all younger VMI men - "it's not like it was back in the 'Old Corps!'"

He was Major General John Dillard, the highest-ranking Engineer officer in Vietnam - a veteran of World War II with twenty-eight years of service in the Army. He stayed for a little while and I showed him around the project. Then he boarded his chopper and took off. I was to see him again, for he flew onto our road section several times. He always remembered my name, and we would always talk about the good old days at the good old Virginia Military Institute. He won my heart.

At that time there were still a few World War II veterans on active duty – all senior ranking officers and NCOs. They had about them an aura of another time – a time in America of self-assurance and unity. It was a time I remember from my earliest recollections, for I was born the year that war ended, and as I was coming along I heard all the old stories from around the dinner table from my Parents, my Grandparents, and my Uncles and Aunts, about war in Burma, war in Europe, war at sea in the North Atlantic, and war on the Home Front. The whole country was in it together. From the radio on the kitchen counter I grew up listening to the music of the Big Band era, and at the movies you could see President Truman on the newsreels. One day we all piled into my Grandfather's car and went down to the railroad station to see General Eisenhower. He came through Lynchburg on a whistle stop when he was running for President, and he spoke to the crowd from the platform on the car at the rear of the train. I felt I was living in the afterglow of a glorious era. And then to stand up there on the road under the command of General Dillard, a veteran of that war and still in uniform, gave me a sense of continuity with something in America that I felt had almost slipped away. General Dillard was like the Fisher-King – the keeper of the Grail…..

As the project progressed, everything moved along eastward to the vicinity of the village. Dennis Casazza had passed the bypass junction with his lay-down site, Jeff Meislin had begun construction on the big two-pipe culvert at the western end of the village by the "monsoon mud-hole," and the old borrow pit by the water point had been abandoned for a new one closer to the lay down site.

The water point was left back down the road in isolation. It had always been in a vulnerable location, but it was the only place that was accessible for the water trucks. Both this creek and the swamp south of the road across from the village afforded the Viet Cong good operational access around our road project. A number of times in the evening we would sit upon the ramparts of the fire base and watch Cobra gunships working over the area around our water point with Aerial Rocket Artillery (ARA). We also would find left-over unexploded ordinance in our first laterite pit that we would have to blow with demolitions, so we felt that there was enough VC activity around there to cause particular concern.

At the water point – which was at the back of the old pit – there was a big mound of dirt. There the guard would usually post himself with his machine gun. The mound was hard-packed, but still I worried about Charlie sticking a mine in there somewhere. To the east of the water point was a berm shoved up at the edge of the clearing that the Rome Plows had made when they cleared the rubber plantation away from the road. The berm ran eastward away from the water point, and along the south side of the creek. Across the creek was scrub brush that had not been cleared, and beyond that was a tree line.

While leaving the big culvert site by the village one day, heading west towards the borrow pit, I got a call on the radio from the guard at the water point: “Sir! They’re shootin’ at us!” I had the driver punch the jeep ahead as I replied: “Well shoot back!”

As we headed rapidly towards the water point I called everyone on the radio and told them to shut everything down, get their rifles, and get everyone over to the water point as quickly as they could, any way that they could. Meanwhile the M-60 at the water point started cutting up the bushes across the creek.

If this had been a diversionary tactic on Charlie's part, he could have drawn us off like this, slipped into the project sites, and blown up a bunch of our equipment while we were gone. But the VC had never exhibited towards us anywhere near that level of sophistication in his tactics. I did not know how much fire the men at the water point were receiving, but my instinct at the time was to do like Stonewall Jackson did at Cedar Mountain – which was to march to the sound of the guns and open the fight with the front file of fours. Any fight is a risk. You just have to try to make it a calculated one, and not shilly-shally around with things once it gets started. Besides, we were all getting a little tired of the way Charlie kept pestering the hell out of us while we were trying to build this blessed road.

As the M-60 was chattering away, the Road Gang came pouring down the road to the water point just like – well, not just like the U. S. Cavalry exactly, but more like the motliest bunch of Mounted Infantry you ever saw. Turning to the right to keep behind the berm for cover, they dismounted off of jeeps, three-quarter ton trucks, dump trucks, low boy trucks, off of you name it from the construction sites. They lined up along the top of the berm and cut loose with their M-16s, tearing up the scrub brush on the other side of the creek. It was a glorious firefight.

After awhile, we gave the "Cease fire" and took stock of the situation. All was quiet across the creek. I don't think Charlie had expected to stir up such a hornet's nest, and when he did, he discovered rather more rapidly than otherwise that he had bitten off quite a bit more than he could chew.

I didn't think it would be appropriate for me to stay behind the berm and call over to the men at the water point that all was clear, so I walked on over there across the open space where the road came in to tell them so in person. It was. By then Charlie was long gone *didi mao* back to wherever his comrades were waiting for him. "Boys," he would be telling them, "don't mess with the Road Gang!"

(l-r) Sgt. Stiggers, Sgt. Cruz, Lt. Meislin
at 2nd Laterite Pit

CHAPTER VI

FRIENDLY FIRE

That night we built the Road Gang's answer to the Rome Plow. The two Maintenance men, a dozer operator, and I brought a bulldozer into the maintenance tent and worked all night long with a welder and a cutting torch fabricating a protective enclosure around the operator's seat. We took airfield landing mats and made a hollow revetment about a foot thick to cover sides and back, brought it up to about shoulder level to a seated operator, and filled it up with sandbags. To top off our "Plow" with just the right touch, we took red lead paint, and with a four inch brush wrote "**** CHARLIE" in great big letters across the blade. Then, with the sun coming up, we loaded it onto a lowboy. After a good working-man's breakfast - courtesy of the chefs of the Road Gang's fine culinary department - we went out with the rest of the equipment that morning and hauled the rig down to the water point to clear off the scrub brush from across the creek.

I got up in the cab alongside of the operator to ride "shotgun" for him while he cut the trace to mark the boundary of the area to be cleared. After we had marked out a sizeable area for the clearing to be done, I dismounted and put a rifleman up there in my place, to cover for the operator while he worked the bulldozer. The guard with the machine gun was also at his post across the creek. The hope was, of course, that this clearing would help preclude the Viet Cong from making another attempt on the isolated water point, by depriving him of the concealment it had afforded him so close to our operation there. We should have cleared it off long before. My mistake.

As the bulldozer started back to work, I saw a jeep pull in to the water point with the Battalion S-3, or Operations Officer. The S-3 was the overall Project Superintendent of the Battalion, acting as such in the name of the Battalion Commander. During the time I was with the 34th, we had three consecutive Operations Officers. They normally held the rank of Major or – as in this case - senior Captain. The S-3 was directly involved with the projects, and would go so far as to assign projects through a Company down to a Platoon. His office calculated materials and made up the construction schedules for the different jobs, and the companies had to submit regular progress reports to the S-3. The Company Commanders in regards to the road project were analogous to Highway Foremen in civilian life answering to the Project Superintendent.

This particular Operations Officer was the second S-3 we had had since I had been there. He was sort of a lean and wiry type, like a kind of fellow that you might run into in a roadhouse someplace out there in the mountains of southwest Virginia, and who looks like he might have a certain familiarity with a Barlowe knife. As I crossed back over the creek, he got out of the jeep and up in my face, wanting to know what I thought I was doing riding around on that bulldozer in the bushes while we were behind schedule on our culverts up on the road! I told him people were shooting at us from those bushes and we were going to clear them off, and I was showing the operator how much clearing to do. As I say, this latest S-3 was a Captain, but senior to me in grade and position, and I wasn't too sure how much I could give back to him. But my year in the "Rat Line" at VMI had finally succeeded in pounding a good lesson into my hard head: "When in doubt, keep your mouth shut"; and when he went on about the culverts I refrained from telling him where I'd like to install the next one.

I had been getting pressure about the progress of the culvert construction before. I explained what I considered to be the demands of security, and how my priority of keeping all the equipment operating caused the burden of security to fall more heavily on the Construction Platoon that had the culvert jobs. I was never specifically told by anyone to lay off the security thing, nor did I ever expect anyone to tell me to do so, but I got the distinct impression that in some quarters at battalion level it was considered that I was making a mountain out of a mole hill at the expense of the construction schedule. A subsequent event some little time later on was to prove I was not. However, at this time I could feel that unspoken pressure mounting, but I never gave in to it. It is sometimes difficult to gage the effectiveness of preventive measures by what has not occurred, and I am not discounting the fortunes of war, but I don't have to live with the ghosts of any of my men. Not all who served in the 34th can say that.

But the incident with the S-3 at the water point was only the latest battle in my private war with the battalion. It had begun early on with some desultory skirmishing. The first shots were fired on Christmas morning, as previously mentioned, with my creative interpretation of orders to include "equipment maintenance" within the definition of "work." I lost that first engagement, and rightfully so, but it did not deter me from trying it again later on – with the same results.

But meanwhile another conflict had arisen from a seemingly harmless set of circumstances. I thought it would be a good thing for us to proclaim our *esprit de corps* by stenciling our moniker on the radiator cowlings of all the company vehicles and equipment. Fighter pilots did it. Other outfits did it. We would see convoys come up the road with it on their trucks: tankers going by with "Petro-Main" on the cowling, or a convoy from some Engineer Dump Truck Company proclaiming "We Drop Our Rocks Anywhere." We were Delta Company, and eating dust, so we naturally were "Dusty Delta." We made up stencils and spray-painted it on all the company vehicles and equipment. That didn't last but about a week, when the Colonel told me to get rid of all that. Of course I complied. That in itself would have been the end of it, and no big deal. But I had to go and have another bright idea.

As Rhett Butler once said to Scarlett O'Hara: "It appears that I have a weakness for Lost Causes." Somehow I had heard that the Viet Cong were superstitious about the Ace of Spades. To play on this, I had heard that some grunts after a firefight would take an Ace of Spades as a "calling card" and hammer it into the forehead of a dead Charlie, using a cartridge for the nail. I can't vouch for the veracity of that report, but I thought it would be a good thing to tape the Ace of Spades to my jeep, just below the center of thc windshield. This lasted a day or two until the card came unglued. Of course I then had to go with the brilliant idea of painting the Ace of Spades where Dusty Delta had previously been stenciled. I suppose it would have been natural for the Colonel to consider this to be splitting hairs with insubordination, but it was in no way intended to be. But I really did have better things to be doing, and the upshot of it all was that I was told to paint over that also, and not put anything of any sort in its place or any other place remotely like it. In short, I was to give up on all notions of that nature whatsoever! I could tell right away that I wasn't in the 82nd Airborne anymore – and, looking back, I think perhaps in my service with that rollicking bunch I had developed some habits that would subsequently be detrimental to my future well-being in my service with the stodgy old Engineers.

The next time I crossed swords with the Colonel was along about February. The battalion notified us that a Huey was being sent up to the fire base for the purpose of giving the men who so desired a helicopter ride over the road project, so that they might have the chance to look at things from a little different perspective. It was a helicopter war, and it would give the unsung Engineers a chance to say they had been up in a chopper at least once while they were in Vietnam. Delta and Charlie Companies were each given two hours one morning for this excursion. We flipped a coin to see who would go first, and C Company won.

We were told that while the first company's personnel were flying around, the other company was to go out on the road and begin work. I thought that since I was not going to leave all the equipment sitting out there on the project sites unattended for two hours while the men came back in for helicopter rides, that it would all be a waste of valuable time to go out in the first place. As soon we got everybody out there, unloaded the dozers and the rest of the equipment and started to work, it would be time to load everything back up again and come back in. We also were in bad need of maintenance on a lot of the equipment. Therefore I disobeyed the order and kept the men at the fire base pulling the needed maintenance while we (I did not take a ride myself) waited our turn for the helicopter rides. I figured that by the time I got called into the Colonel's office over that one it would all have been *fait accompli*, and that I would go ahead and take my medicine like a man and suffer for a noble cause.

I guess I could have gotten court-martialed for insubordination. In the Prussian Army I would have. In the Chinese Army the Colonel would have just pulled out his pistol and shot me. But the long-suffering Colonel of the 34th Engineers was satisfied with calling me into his office once again and "explaining" things. Once again, he was right. The breakdown of discipline outweighs the breakdown of equipment.

But not disputing that, any maintenance done pays dividends. If you don't take the time for maintenance when you have the chance, the equipment will take it when it wants to – usually at the worst possible time. I always found that it is easier to keep something running with good maintenance, than it is to try and get something running again from a lack of good maintenance.

My next conflict with battalion was over an incident involving the maintenance of a scraper tractor. A Sergeant from the Equipment Platoon came up to me one day with one of the scraper operators – who looked worried sick. I was told that the engine had seized up on the man's scraper and the man was going to be court-martialed for running it without oil in it. The operator said that when he had changed the oil the night before, he had put new oil back into the engine afterwards, and his NCO testified that he watched him do so. They said that the oil pump had gone out and that is why the engine seized up. I asked why, then, was he going to be court-martialed? They told me that solidified babbitt from the crank bearings had been found stuck to the bottom of the oil pan. The people at battalion were saying that if there had been any oil in the crankcase at the time, there could not have been any babbitt stuck to the oil pan. The operator was being accused of burning up the engine, and then putting oil into the crankcase to cover up his misdeed. But both the operator and the Sergeant were earnest in their assurances to me that the operator had not run the engine without oil in it.

I knew both to be good men, and I had no doubt that they were telling me the truth. I told the operator I would try and get him off the hook if I could figure out how the babbitt could have gotten stuck to the oil pan with oil still in the crankcase. It would be a problem of Thermodynamics, and unfortunately Thermodynamics had been a problem for me. However, I had taken the course twice, so I reasoned that I should therefore know twice as much about it as the average Engineer. Furthermore, when one must solve a problem for someone else who is depending on him to get the right answer, one is far more inspired to pay attention to the subject than is the scholar who is sitting in summer school on his own behalf, with the birds chirping in the sunshine outside, and with a dear little Randolph-Macon girl on his mind.

I went down to Phu Loi and locked myself in my office in the orderly room for two days. I believed what the men had told me, so there was bound to be an explanation. What I finally came up with was this: that the solution hinged upon the fact that once a solid material has melted, the temperature of the resulting liquid can continue to climb. This is obvious if you think in terms of water. Solid water (ice) becomes liquid water at 32 degrees, and if it continues to be heated, liquid water will rise in temperature even to the boiling point of 212 degrees - where it becomes gaseous water (steam). This basic principle applies to babbitt just as it does to water, or any other material. The only differences are the actual melting and boiling points, which are irrelevant to the argument.

The application here was, that if there was oil in the crankcase as claimed by the operator, and if the oil pump had gone out on the engine - thereby keeping the oil from circulating and lubricating the engine - then heat of friction would have been generated. This heat would have continued to rise until it started to melt the babbitt in the bearings, causing the babbitt to be flung out of the bearings and into the oil as the crankshaft continued to spin.

The first of the babbitt to be flung out would have been that which was exposed on the outer edges of the bearings. It would have been flung as soon as it reached its melting temperature. Chances are that this babbitt would have been cooled enough when it first hit the oil to bead up and solidify before it got through the oil to the bottom of the pan - sort of like making old fashioned "drop-shot" for shotgun shells.

But while this babbitt was being flung out, the remainder of the babbitt in the interior portion of the bearings would still be absorbing the heat of friction, as the engine would not have seized up until all of the babbitt was gone. Some of this babbitt could have reached a temperature high enough above its melting point - before it was flung from the bearings - to allow it to pass through the oil and stick to the oil pan before the oil had a chance to cool it back down enough for it to become solidified again.

It was the best argument I could come up with. I presented the argument to the Colonel, who accepted it as giving reasonable doubt, and it was enough to get the operator off the hook. It was also enough, I would later come to believe, to get me marked by someone in the battalion who had "lost face" over the issue as a result of my argument - and causing someone to lose face is not a good thing to do in the Orient.

Maybe I was just being paranoid. But a situation of a different sort cropped up soon afterwards that caused Battalion Operations to lose face in a big way. It involved a major error in their production schedule for my road section. It involved me, because I was the one to point it out - and not in the most diplomatic fashion possible.

At the beginning of the Highway 1A/2A project, the S-3 had calculated a construction schedule for each company based primarily on the volume of material that was required to complete the sub-grade on each road section, and the estimated amount per day that could be put down. I was given a copy of this schedule at the beginning of construction. I was told to keep a daily tally of the amount of material that we put down and to submit this tally on a regular basis to Operations. Each scraper carried seventeen cubic yards of material, and all that the Sergeant at the lay-down site had to do was to make a mark on the tally sheet each time a pan put down a load. I always had a running total, and S-3 always had a running total as well.

At some point I started getting pressure not only about the culverts, but about the fact that we were not as far along up the road as the construction schedule projected that we should have been. I took another look at the schedule and then another look at the calculations that Operations had based this schedule upon. I discovered by my running tally that we had already put down the *total* amount of laterite that S-3 had originally calculated would be required for our *entire road section*, yet we weren't but about halfway along on the section! It was a big mistake in S-3's calculation of the amount of material required to complete that portion of the road! Such a fundamental mistake in the calculations for the schedule could have caused the battalion to fail in carrying out its primary mission before the monsoon, because it could have caused the battalion to become unknowingly committed beyond its capabilities. This was not a good thing.

Perhaps I should not have accepted the construction schedule at face value in the first place, but it was only after being pressured for being behind that I looked at the schedule critically and found the error. This I reported immediately to the Colonel, without any malice towards anybody in particular. But I confess it gave me a distinct sense of vindication in general for all the guff I had been getting for supposedly being behind schedule.

If it vindicated me, it made the S-3 look bad. Had I looked for and found the error sooner - and gone straight to the S-3 with it instead of to the Colonel - Operations would probably have done some re-calculations, I would maybe have been praised for being ahead instead of harassed for being behind schedule, and all would have been well. But I was not a seasoned practitioner of inter-office diplomacy, although I should have been; much of that is just basic consideration for others. If I had used a little more tact to try and resolve the situation, I would have not only made myself look better, but I could have made myself a loyal friend of the S-3 with little or no effort. Friends are good things to have, and I could have used one a little later on.

I did none of those things. When I carried this little tidbit of information back down to the Colonel, he had Operations revise the project schedule. When it was re-calculated, it was determined that the project would require about *twice* the amount of material on D Company's road section than that which had originally been calculated. Based on the revised calculations, therefore, the Road Gang was not behind, but way ahead of schedule; and it looked like we would still beat the monsoon if nothing happened to change the way we had been doing things.

But something did change things. Just about the time we were congratulating ourselves for being ahead of schedule, our Clark 290 scrapers were taken away from us, and we were given Caterpillar 830s to take their place. This – like the M-16 rifle - was evidently another piece of equipment that some pork-barrel Senator had awarded a government contract on to one of his constituents. If the middle of a firefight is not the time to be field-testing new rifles, then the middle of a race to beat the monsoon is not the time to be trying out a new piece of construction equipment – especially if that piece of equipment happens to be the one that decides whether or not you are going to beat the monsoon. Give the thing to a civilian contractor free of charge and let him work out all the bugs before you dump it on somebody who is trying to fight a war. As if we didn't have enough troubles.

The 830s had this pipe elbow coming out of the hydraulic oil tank that could not stand the vibration of normal operation. After a little while it would crack and start spraying oil. We had a few spares in inventory, but they were soon used up. Since it was a new piece of equipment that we were supposed to be field-testing, there weren't too many spare parts floating around in Army inventory. And here is an example of why governments cannot compete with the private sector for survival: government bureaucrats don't have to produce a profit or a success – or even win a war – to keep their jobs, as long as they keep the paperwork in order. As long as we put that part on requisition, we could pass all the Inspector General inspections in the world and get fabulous fitness reports for doing so, but without that part we couldn't get the road built. The new 830s wouldn't run, and nobody would give us back our 290s, yet they expected the road to be built on time! Like the Israelites in Egypt, the Road Gang found itself tasked with making bricks without straw.

But as we said, "Please be patient. The difficult we do immediately; the impossible may take a little longer."

Whenever a pan went on "deadline" for any other repairs, we took the elbow off that tractor and put it on one that only lacked that elbow to be made operational. Occasionally our parts people at Phu Loi would receive one and get it up to us, but at best we dropped back from having an average of six pans running all the time, to an average of three or four. The sudden requirement in the middle of the project to double production with half the means knocked us back pretty badly. However, with our previously gained advantage of being ahead of schedule, we had a large enough lead on the material to allow us to keep ahead for awhile – although the schedule started gaining on us with no relief in sight. We hoped we would be able to stay ahead long enough to complete our section before the monsoon hit in May. It looked like we could, but it would be close, and there would be pressure.

Other pressures began to manifest themselves besides those related to the construction schedule. One night there was a commotion over in Charlie Company. The Commander had shot himself in the fleshy part of his thigh with his .45 pistol. He said it went off while he was cleaning it. A Med-Evac helicopter had been called in, and was on the way, and the pilot had requested some illumination for an LZ. We rounded up some jeeps and three-quarter ton trucks and turned the headlights onto a place outside of the wire for the chopper to land. I helped to get the Captain out of his bunker and aboard the helicopter, and we didn't see him again after that.

A new Commander for Charlie Company was then sent up with orders that he was to be in command of the fire base. Prior to this I had sort of nominal command due to the fact that the Road Gang had built it and had seniority of occupation I suppose, but when Charlie Company arrived I certainly did not try to exercise control over them – nor was there any need to. If there was an issue to be addressed, the C Company Commander and I just got together on it and that was that. They generally adopted our security arrangements within the fire base and took care of their side, and we took care of ours. They had their own entrance gate and they came and went any way that they saw fit. Units that came in to stay at the fire base for awhile usually had an officer who outranked us both, but except for the 1st Infantry curtailing the beer ration while they were there, everybody generally tended to their own people and we tended to ours. When the new C Company Commander arrived things continued just as before, but I gathered from circumstances aforementioned that these new orders had been sent from Battalion in part as a rebuke to me.

Some time later an Armored Cavalry unit came in with their Armored Personnel Carriers (APCs) and stayed for a week or so. I asked the Commander if they would cover us while we went out and looked for some laterite closer to the lay down site. This would increase production and help make up for our shortage of scrapers by shortening their turn-around times. The Cavalry Commander agreed, and I rode perched upon one of their tracks as they escorted one of our exploratory bulldozers, but we found nothing worth excavating.

A day or two later, as I was on Lieutenant Meislin's culvert site, we watched the Armored Cav engage Charlie over in the same location where we had been looking for laterite a few days before. Airmobile Infantry was called in, and they dismounted from their choppers and made a sweep of the area, but they found no sign of the Viet Cong.

That Cav unit was the most demoralized unit I ever saw during my tour. At night they would have a party in their APCs. I heard they would sometimes sneak "boom-boom" girls from the village into their tracks. You could sometimes catch a whiff of pot in the air. I was told that dope, which had become a big problem in some units by then, could be bought in a bag the size of a pillow case for about fourteen dollars. It was one sign among many that the U. S. Army was becoming demoralized.

With the Vietnamization of the war, America was getting out – but it was still drafting men and sending them over to fight in a war that nobody cared about anymore. Replacements going over to Vietnam were meeting troop units - and body bags - coming back home. As Stanley Karnow said in his *Vietnam: A History:* nobody wanted to be the last man to die in a war that had lost its meaning. The replacements were arriving demoralized, and the infection was beginning to spread.

It was evident that these Armored Cavalry troopers were out of control. One day I came up from Phu Loi and was told that Lieutenant Meislin had been "fragged." I suspected that someone from our company had gotten mixed up with some of that bunch and had been grumbling about Jeff. Grumbling is normal in the Army, but with that crowd it went too far very fast, and somebody just flipped a fragmentation (hence the term) grenade into his bunker one night. Luckily it lodged on the sandbags outside, instead of rolling down into it and under his cot. It went off and blew holes in the Medic tent, but no one was hurt. There was no way of finding out who did it.

"Fragging" was this lovely improvement on the French Revolution. If you got conscripted and didn't like the Vietnam War, just take it out on the nearest person in authority. Don't worry if it was a war that "Camelot" Kennedy and "Great Society" Johnson got us into, and don't worry if the Officer or NCO was just doing his

Duty and didn't like it any better than you did. Lots of NCOs and junior officers got a Purple Heart and a body bag from this "friendly fire." Some Aquarian conscript high on dope would just stroll by and roll a hand grenade into your hooch.

A couple of years later the United States discontinued the draft and went to an all-volunteer Army. Revolution accomplished. Now the enemy is us.

Those Cav people didn't stay with us for long, and I was glad to see them go. I'd rather deal with the Viet Cong without their help, thank you very much! They exercised a bad influence on some of our men. One evening soon after their departure I pulled into the fire base in the jeep and saw a large group of men standing in a ring that extended up onto the parapet by the southeastern bunker. In the middle of the ring was one of our men brandishing his M-16 and running around like he was crazy. The men surrounding him kept at a healthy distance, yet they were drawn to the spectacle. The ring of men looked like the ripple from a rock that had just been thrown into a pond.

It is times like these that take all the fun out of being the Captain. I had to do something, so I made a little show of unbuckling my pistol belt and handing it to a Sergeant standing nearby. Then I walked up to the man with my hand out and told him to give me the rifle, which he did.

From what I gathered afterwards, the man had fallen in love with some girl in the village, who had messed up his mind. Well, I know how that is, but I couldn't have everybody going *dinka dao* on me at this stage of the game. I gave the man the maximum a Company Commander could give under Article 15 of the Uniform Code of Military Justice (UCMJ): reduction one grade in rank, with forfeiture of one month's pay. I also sent him back to Phu Loi to get him away from this woman who was driving him crazy. He did just fine after that. No one in his right mind can advocate mixing men and women in the same military unit – unless his agenda is to undermine and demoralize it.

But all in all we were very fortunate to have maintained good morale in the Engineers. The objectives of the war – which had started out fuzzy – had devolved into near meaninglessness. Objectives of many units had likewise devolved. Men would go out on sweeps or ambush patrols merely for the purpose of racking up body counts. It was like violently sloshing about trying to shove water away, only to have it flow back again as soon as you withdrew from the exercise. There would be a successful ambush, some Viet Cong would get zapped, some American soldier might trip a booby trap during the exfiltration and have his legs blown off, and then the vast Asian tide of humanity would flow back in and fill up the gap left by that violent little action as though nothing had ever happened. And nobody cared except the men who had to go out and do it again.

But the Road Gang had a clear-cut objective, and we could see the evidence of our progress towards it every day - evidence that was still there the next morning when we started work again. It was the very best thing for morale, for as the project got closer and closer to Claymore Corners, each day would give us a little victory for our efforts - even if we were only building a road to nowhere.

(l-r) Sgt. Baker, Lt. Casazza, Lt. Stienbeck
on Bypass Road

CHAPTER VII

BRICKS WITHOUT STRAW

By the first of April things were coming down to the wire. The monsoon rains would be starting in a few weeks, and the S-3's revised Project Schedule – which had started gaining on us the day that our Clark 290 scrapers had been replaced with the faulty Caterpillar 830s – had finally caught us. If something could not be done, from this point on we would start falling farther behind as each day passed. We still were only able to keep about half of our scrapers hauling, due to the failure of the hydraulic elbows, and there would be no improvement in that situation for the rest of the time I was there. The only way I could think of to increase our production was to find another source of laterite closer to the lay down site.

By this time we were within two kilometers of Claymore Corners. Lt. Meislin was working on the last culvert site, which was the big double culvert at the "monsoon mud-hole." Lt. Casazza had come through that low section of our project with the lay down site, and was now at the eastern end of the village putting down dirt. Again we had gotten to higher ground which would require less material. This should have sped up the linear progress on the road, but we were working progressively farther and farther away from our pit. This made for longer and longer turn around times for the scrapers, and negated the advantages of less material requirements. With only half the scrapers able to operate, it was going to be a tight race to beat the rains. If we could find laterite to the east of us, we would then be working towards the pit instead of away from it. Thus we would pick up speed as we approached the end of the project, and hopefully at least break even on the schedule. As it was now, we were slowing down the closer we got to the end.

The only place that we had not explored was east of Claymore Corners. Beyond the intersection there had been a strip of rubber plantation along Highway 1A that had been cleared off in the past. This had grown back in low scrub. About a half-kilometer to the east of 1A was a creek that flowed north around a knoll before turning east into a swamp that ran through a large section of forest. I thought we might look between the road and the creek, and if necessary, explore beyond it on the knoll. We could roll a culvert into the creek bottom if we had to go over to the other side.

I had not forgotten how Claymore Corners had gotten its name, and I knew the terrain had favored Viet Cong operations in the past. There had been a convoy war going on in III Corps within recent times. Charlie had found the convoys good hunting when their escorts had gotten stretched thin, and this was one of the ambush sites. The situation had improved for us, and I didn't know of any specific activity going on over there east of the intersection at the time, either friendly or otherwise, but I knew there was nothing over there but pure "boonies" – and going hunting over there would not exactly be like going squirrel hunting over on Mr. Catherwood's farm. I decided that we ought to take along some of our men as "Mounted Infantry" in a couple of trucks to act as security for the bulldozer while we were scratching around over there. I would go ahead with a bob-tailed tractor and push brush for booby traps or whatever as we checked out some locations.

I remember waking up in my bunker about 3 A.M. worrying about it. When we got underway in the morning, I took some filters out of a couple of my cigarettes and poked them in my ears, so that if I did hit something maybe the explosion wouldn't damage my eardrums.

We pushed along and scratched around with the dozer but did not find anything suitable between the road and the creek. There was a trail that looked like a cart track leading across the creek towards the knoll and we started along that, with me pushing brush and the little security convoy trailing behind with the bulldozer. When I got to the creek I saw that the track made a dip down into the bottom of a small gully and then up the other side. To stabilize the crossing of the creek itself, someone had rolled a bunch of unexploded artillery ordinance into it as ballast. It looked as if it was well settled and had been run over a number of times in the past, so I took a chance with U. S. Government property and crossed over.

Looking back I probably should have devised another plan – and we should have gone back with demolitions and gotten rid of them - but after being "in country" awhile one begins to say to oneself: "Oh, well, 90 percent chance that it will be OK. Good enough." It is the sign of a dangerous complacency setting in - there was enough explosive there to launch the first manned scraper tractor into orbit, courtesy of the U. S. Army Corps of Engineers. NASA would have been green with envy. But after all we found no more laterite, so our existing pit had to serve us for the remainder of our project.

After our little fight at the water point, Charlie started trying his luck with C Company up highway 1A – "Charlie" vs. "Charlie." Nowhere was the danger of predictability more sadly exemplified than on their lay down site, where evidently they were in the habit of setting up a water cooler in the same spot every day. A truck driver stopped, walked over to get a drink of water, stepped on the mine that the VC had planted there, and was literally blown to bits.

Later we heard that the C Company Commander and his driver had just missed being hit by a command-detonated mine. Had Charlie detonated the mine just ahead of the jeep they would have had as much success as if they had timed it perfectly, so it had to have gone off behind the jeep. They don't do much bird hunting in Vietnam, evidently, so the VC weren't always proficient in judging lead. U. S. helicopters were the only birds they had to practice on, but that was not a very healthy way to polish those skills. In that case one was more likely to become deceased before one became proficient.

We were happy for their deficiency. Each month I would offer the job of jeep driver to someone who had been recommended to me as deserving some recognition for his good effort. I had thought it might be a welcome relief from the dust of the road, but after hearing the snap of a bullet behind my ears a time or two I concluded that a jeep with a radio might be considered a nice target to practice lead on – and I wondered if that sort of recognition was considered by the troops to be such a desirable reward after all.

Besides having our dirt-hauling capabilities cut almost in half because of the maintenance problem with the 830s, another chronic problem appeared. We had two sheep's-foot rollers in TO&E. One roller was used on the road at the lay-down site, and the other was used to compact the slope on the shoulders of the road. The problem developed that the tongue on one kept breaking. Our two miracle-workers in the Maintenance Section would build it up again with weld at night, and it would hold for a little while, but then it would break again. The priority was the lay-down site, so the problem roller was relegated to the shoulder compaction, and we necessarily began to fall farther behind on that aspect of the job as well – another deficiency to add to my list.

Not all horse thieves are Supply Sergeants, and not all Supply Sergeants are horse thieves, but there are similar affinities for appropriation to be found among some practitioners of either trade. My good Supply Sergeant was apprised of our plight over the defective sheep's foot roller and had done what he could through channels to get us a new one. Of course there were requisitional procedures to follow, and channels to go through with battalion, brigade, etc., etc. You can imagine how this was going to help us beat the monsoon. So the Sergeant contacted some of his friends (all Supply Sergeants have lots of friends) and found a two-drum sheep's foot roller that somebody had down at Long Binh. All we had to do was to send a truck down there and pick it up.

Security requirements had kept us behind on the culverts throughout the project, the faulty hydraulic pipe elbows on the scrapers had lately put us behind on the material at the lay down site, and now a disfunctional sheep's foot roller was putting us further behind on the shoulder compaction. I thought this wonderful discovery of a free roller would be happily received by battalion, who had been dunning me all this time about the schedule. Our problems would be over in that regard at least.

I had finally learned by then that I had better not make any more unilateral off-the-record-type decisions of that nature without first clearing them with Headquarters, so I went to see the Colonel to ask his permission before I went down to pick it up. The Colonel was not in, so I went to the Executive Officer and asked him. He told me No, the roller was not authorized in our TO&E, and we couldn't have it. I asked him if I could take our bad one down and trade it for the good one - or just dump it off somewhere so that we would be in compliance with the TO&E - and then get on with the war? No, I could not. I was flabbergasted! Just how badly did anybody want this road built anyway?

The XO was a pretty cranky man. Somebody told me later he had gone on R&R to Hawaii to meet his wife and she didn't show up. I guess that would make anybody cranky, but I didn't know if he was a grouch because she didn't show up, or she didn't show up because he was a grouch. I don't feel compelled to meddle in other people's personal problems - be they the Major's marital relations, the race relations in Detroit, or the political relations between North and South Vietnam - but sometimes you get caught up in them. All I could do was thank the Supply Sergeant for going the extra mile, tell Junior to keep plenty of welding rods on hand up there at the fire base, and keep on making bricks without straw.

Of course, somewhere along the line I flunked an IG inspection, so that might have been a contributing factor in the XO's attitude about the TO&E. But I think the inspection was afterwards, because that deficiency was not called to my attention in the interview, and I am sure the Major would have reminded me of it otherwise.

My good old First Sergeant Benavente had recently completed his tour and gone home, or I am sure I would have passed the IG inspection. His replacement was not so hot. After awhile I started getting reports that he was disappearing into town somewhere and leaving Lt. Stienbeck holding the bag with everything. I should have done more about it and done it sooner, but to be straightforward, I felt that we had more important things to worry about. At that particular moment I felt that an IG inspection bordered on harassment. There is a time and a place for everything, and I felt that this was not the time for all that. Wait until after we beat the monsoon and then have all the inspections you want. Of course, this was not the proper attitude – but it sort of fit right in with some of the other improper attitudes that I was beginning to develop by that time. Anyway, I flunked the inspection, didn't care very much, and then went back up to the fire base to try and help get the road finished.

At one point the S-3 recommended that I spend a little more time at Phu Loi and a little less time at Fire Base Lobo. Maybe I should have. Jeff and Dennis were up there with the best NCOs anybody could ask for, and all concerned were hustling and doing a great job with what they had. Maybe I was getting in the way. I had been splitting it up with about two or three days at the fire base and one or two days spent at the base camp. I know the Platoon Leaders certainly didn't need me up there underfoot all the time, but the truth is, I would rather have been at Fire Base Lobo than at Phu Loi any day. I remember one time wishing that I was a Platoon Leader again. It is the best job in the Army. I hope I didn't spoil it for Lt. Meislin and Lt. Casazza.

I was told that there was a pinochle game at the club on Tuesday and Thursday evenings that maybe I should join. I think the XO was a regular - and maybe the Colonel - along with some of the Company Commanders and Battalion Staff. Perhaps I should have made an effort, but what I really should have done was communicate more. It would have saved me a lot of grief. I did not confer often with the S-3 other than to supply his office with requested information, such as the material tally, and looking back on it through the years, I was never once called upon to attend any sort of regular battalion briefings during my entire service as Company Commander. I would go and report to the Colonel whenever I came down to Phu Loi, but I do not have any particular recollection of him ever asking this young man to open up and fill him in on problems we might be having, or – if he did - offering me much in the way of suggestions or encouragement in dealing with them. Nor did I really ask it of him. Perhaps we can both share a little blame in that. But as I look back upon it, he was a young man, too, and he had the same problems a Company Commander had - only multiplied by five.

In the last half of April, three things happened: "Operation Twenty-Five Clicks" was completed, the Commander of the 20th Engineer Brigade was promoted, and the monsoon rains began.

The 34th Engineer Battalion was part of the 20th Engineer Brigade, whose Headquarters were located at Binh Thuy, down in the Mekong Delta just upriver from Can Tho. The Commander was a full Colonel, and was to receive his promotion to Brigadier General before the end of the month. In honor of this, the brigade was to make a push to pave a total of twenty-five kilometers of their various road projects in the week that preceded the Commander's upcoming promotion. When this was done it would all be put on the front page of the Army's newspaper, the Stars and Stripes, along with a picture of the newly promoted Brigadier General of the 20th Engineer Brigade.

The 34th Engineer Battalion's portion of the Operation was to pave seven of those twenty-five kilometers, which would begin below our road section on Highway 2A. The project would not affect D Company particularly, except that the last two or three kilometers of pavement would extend onto our road section – all of which had been completed and was ready for the paving operation. The end of the pavement would be somewhere on our section between the bend in the road where it turned towards the east, and the turnout to the laterite pit. When B Company started putting down the rock base course for the asphalt at the beginning of the seven kilometers to be paved, we were over ten kilometers up the road ahead of them with the sub-grade, and within sight of Claymore Corners. We got our first rain shower that afternoon, foretelling the arrival of the southwest monsoon.

During the week of this special project, the Brigade Commander was going around to the different battalions and inspecting the various road projects. I was directed to prepare a presentation for when he arrived on our job site, in order to inform him of our progress. Our Colonel was to escort the Brigade Commander and would also be present. I had a couple of days' notice prior to his arrival. All I had to do was set up the easel with the Project Schedule that the S-3 had prepared, brief him on our actual progress, and then answer any questions he might have. At that time we had dropped about two weeks behind S-3's revised Project Schedule due to our equipment problems.

The evening before the presentation I was at the fire base sitting on top of Lt. Casazza's bunker discussing it with him. We checked the tally on the material against the S-3's revised Project Schedule, and discovered that once again the Road Gang had surpassed the total amount of material that Operations had calculated for the entire project. Had their estimate been correct, we would have already been finished – even with only half of our scrapers working - yet we were still over a kilometer from the end of the project. I told Dennis that I did not mind explaining to the Brigade Commander our being behind schedule based on the failure of the scrapers and the unavailability of the parts, but I did mind us having to take the blame for S-3 having once again botched the Project Schedule by mis-calculating the material requirements for our road section.

Once again Captain Quixote took the surveys and calculated how much more material it would take to get to Claymore Corners, and how much more time it was going to take to finish the job. If this math will cause the eyes to glaze over, then skip the next paragraph.

I first figured the average time it would take for a scraper to load at the pit, run to the lay down site and dump, and then return to the pit. From this I could figure how many trips per day one scraper could average. Since each pan held seventeen cubic yards of laterite, I could then determine how much material one scraper could put down per day. If we could figure on an average of four scrapers operating, then four times the material calculated for one pan would give us our total daily production rate. Our daily rate divided into the total amount of material required to finish the road would then give us the days needed to complete the project. I figured it all up and concluded that it would take us two more weeks to finish. I based my calculations upon the construction drawings supplied me by S-3.

It has been a third of a century ago, but I think I remember a very good turn around time for a scraper was twenty minutes; but to go with a very conservative estimate of an hour, four pans working ten hours would give a production rate of 680 cubic yards of laterite per day. This is a very conservative estimate, but if Battalion Operations figured that we should have been finished two weeks prior, and we had in fact surpassed their totals and still required two more weeks, then S-3 had mis-calculated material by somewhere between 9, 520 and 19,040 cubic yards.

By the time I had my briefing prepared for the next day, and had done these calculations, it was well after dark. I wasn't going to put U. S. Government property at risk by driving down to Phu Loi in the middle of the night, but I wanted to tell the Colonel beforehand that I was going to present these figures at my briefing, if the Brigade Commander asked me when we were going to complete our project. It would be a routine question that was sure to be asked.

I went to the communications bunker and called up battalion. I asked the radio operator that answered to please let me speak to the Colonel. After a moment he came back and said that the Colonel was unavailable, and he gave me the S-3 instead. I earnestly explained everything to him.

At this date I look back in wonder at how innocent of guile I was. Here I was on the eve of the Brigade Commander's inspection of the battalion's biggest project, telling the Battalion Operations Officer that his material calculations were wrong, that we were not behind as his Project Schedule said we were, that I figured we had to put down fifteen thousand yards of dirt more than he had figured in his calculation, that it would take us two more weeks to do so, and that I was going to use my correct figures instead of his incorrect figures to give our completion date to the Brigade Commander in my briefing tomorrow. I wanted the Colonel to know about it ahead of time so he wouldn't be surprised when I did so, and would the S-3 please be sure and tell the Colonel before he comes up onto the road tomorrow?

Oh, yes indeed, you can depend on me, said the cat to the canary.

You can imagine how that turned out. Back then I was still operating under some quaint notions such as trusting things people told me.

I could point out the spot this very day. I was standing beside my easel next to the lay down site just east of the village. You could see Claymore Corners just beyond the rise. The rain shower of the evening before had wet the road, but the hot tropical sun of the morning had quickly dried it into dust again. The scrapers were putting down laterite while a grader was working it. A dozer was pulling a sheep's foot roller. The dust was beginning to rise in the morning heat, and the air was filled with the incessant drone and growl of diesel engines. Then I saw the Colonel's jeep coming up the road.

I saluted as they got out. The Brigade Commander stood facing my easel with his back to the road. The Colonel was standing to his left and a little behind. I only remember the Brigade Commander asking me two things. The first was how many pans did we have running? I told him, and briefly explained about the hydraulic elbows. His next question was how much longer before we completed the section? I told him two weeks.

I do not know what the Colonel had been telling the Brigade Commander, but Claymore Corners was over a kilometer away, and it was obvious to anyone standing there that we weren't going to finish the road in the next day or two. But I never will forget as long as I live seeing the Colonel's head snap back as if I had thrown a glass of water in his face. The good S-3, of course, had not relayed my message to him the night before. Nothing more was said as they mounted the jeep and drove on up the road to the C Company section – but that was not the end of it by a long shot.

Meanwhile "Operation Twenty-Five Clicks" was merrily rolling along. Company B was putting down the crushed rock base course in about a six inch thickness. Company A was coming along behind and covering it with asphalt. With the rains coming it was important to cover the rock to keep it dry. If the rain soaked in, it would destabilize its compaction and the road surface would not hold up. By the time the paving operation got onto our road section towards the end of the week, the monsoon rain showers were beginning to fall in earnest in the evenings.

Alpha Company could not keep up with their asphalt to cover the rock that Bravo Company was putting down, but the pressure was on to complete the seven kilometers of surfacing before the end of the week - and rains be damned. At the end of the day, there would be a long stretch of rock out in front of the end of the pavement. This was left exposed to the rains, but the next day the wet base course would be paved over. I remember seeing the NCOs shaking their heads as the paving continued.

When you see NCOs shaking their heads, it is not a good thing. I remember driving over the paved portion and seeing the water running out from under the asphalt in rivulets. The rains had soaked the rock and the water was being squeezed out of it like a sponge. It illustrated a classic result of trying to make conditions fit a schedule, instead of making a schedule fit conditions.

But "Operation Twenty-Five Clicks" was completed on time and made the front page of Stars and Stripes. The Brigade Commander got promoted to Brigadier General and got his picture in the paper as well. The directive then went out to all of the Company Commanders in the 34th Engineer Battalion to pick two deserving men from each of their companies and have them present at the end of the pavement the next day. The Colonel would be there and would be awarding them each the Bronze Star for "meritorious service in a combat zone."

I had so many worthy men that I didn't know whom to pick. After thinking about it for awhile, I decided that the Herculean efforts of the two men in the Maintenance Section deserved that recognition. They had lived at the fire base and carried their load alone without a murmur for five long months. Without their efforts we would not have been anywhere near our objective at Claymore Corners. They had kept those scrapers operating on a shoestring, and in between times they were keeping everything else operating on the road that was humanly possible. They were whom I would select, and I knew that there was no man in the Road Gang who would begrudge them for it.

There was only one little problem: the two men did not want any medals.

I tried to explain to them that they deserved medals as much as any man in the company.

Thank you, Sir, but not interested.

I wanted them to have those medals. I asked them to do it. No go. They both just wanted to get back up underneath this bulldozer and finish dropping the oil pan. Finally I had them both hog-tied and rolled into the back seat of the jeep, and we drove down to the end of the pavement where the ceremony was to take place. Hog-tying people takes a little while, so by the time we got there the other men who were to be awarded were already starting to line up on the road; Alpha, Bravo, and Charlie Companies - in order - and then "Dusty Delta" last in line.

Somebody read the citation. Then the Colonel and his attendants started down the line pinning medals and shaking hands with the recipients. After pinning the medals on the men from C Company, they all walked back to their jeeps and drove away. I stood there with my men stunned! I went up to some Staff Officer who had been in attendance and told him they forgot somebody in case they hadn't noticed. He told me they didn't have any more medals.

There is a term prevalent in the Army that describes a deep contempt for pettiness dispensed by someone in authority. In certain circumstances there is no term to be found that is more appropriate - and this was one of those circumstances. It was one big load of pure and unadulterated chicken shit.

This was the Colonel's rebuke to me for the embarrassment I had caused him at the briefing I had given to the Brigade Commander a few days previous. The Colonel had not been told by the S-3 of my material calculations - as I had been assured that he would be - and because of that, the Colonel considered that I had slapped him in the face in front of the Brigade Commander. I had not seen the Colonel in the few days since – still in my innocence thinking that the S-3 had told him of my radio transmission - but I should have gone to him right away anyway and emphasized that I had tried to get in touch with him beforehand; and that I had (naively) left word about it for him with the S-3. That was the least I could - and should - have done. Big oversight number one. But better yet, I should have gone down to Phu Loi well ahead of the Brigade Commander's visit and figured everything up, then gone in to see the Colonel about it personally. This comes from the mellowed perspective of years. To paraphrase the line from Cool Hand Luke, what we had there was a failure to communicate – plus a shortfall in foresight, and in consideration for the Colonel on my part.

But I was not feeling any of that then, standing there on the pavement of 2A. I only felt a great sea-change sweeping over me. I felt a cold contempt for everything. I was not demonstrative about it, but at that moment – deep down - I lost it; I went as *dinka dao* as the man who went nuts that evening at the fire base. I felt that to rebuke me, two good and innocent men of my command had been slapped in the face by a coterie of Phu Loi pinochle players. I did not know for sure who came up with the idea, but I am surprised at the Colonel for indulging in it. I was the one who deserved the rebuke, not those men. It is still the most petty and contemptuous thing I have ever heard of in my life, and it still infuriates me when I think of it….

As I took those two poor embarrassed men back to the fire base, I had a desire to get every man under my command out of that insane asylum and back home in one piece if we never put down another shovel full of laterite on Highway 2A. But the men of the 1st Air Cavalry were depending on us, so we kept on hauling dirt.

A day or two after the Brigade Commander got his promotion and his picture in the paper, the pavement fell apart - just as the wise old NCOs had predicted. Getting too far ahead of the asphalt with the base course and leaving it exposed to the monsoon rains had ruined the compaction, and after a few convoys had run over it the surface began to crumble. *THAT* little sequel of "Operation Twenty-Five Clicks" did not make the front page of Stars and Stripes, but I figured out that day how important the road was to some people – and in what way.

Shortly thereafter I learned that someone at Headquarters was blaming me for the road surface falling apart. I told them to go run compaction tests on our laterite. They did, and they found our foundation to be as solid as a rock. I knew the Road Gang had not let me down.

Culvert Site at the "Monsoon Mud-Hole"
Looking Eastward towards Cau Dinh Village

CHAPTER VIII

THE ROAD TO NOWHERE

There are two great seasons in southern Asia – the dry and the wet. They are each governed by a monsoon, which is the name for a wind. The dry season comes in winter when the great land mass of Asia is cooled. High pressure is then created over the land, while the warmer air mass over the Indian Ocean rises to create a low pressure over the sea. Where the clockwise circulation of the air around the high over the land meets the counterclockwise circulation of the air around the low over the sea – each circulation caused by the Coriolis Effect, due to the rotation of the Earth - a flow of air from the northeast is established. This air flow coming off the land has little moisture, and is called the dry monsoon.

In the summer the cycle is reversed. The Asian land mass warms, creating a low atmospheric pressure above it, while high pressure builds over the relatively cooler Indian Ocean. The clockwise circulation of air is now in the south around the high over the sea, while the counterclockwise circulation is in the north around the low over the land. Thus the air flow where they meet is now from the southwest, and laden with moisture from the Indian Ocean. When this air passes over the warm land mass, it rises and is cooled, causing the moisture in it to condense. This southwest monsoon then dumps its torrential rains and swells into a flood the mighty rivers of Asia: the Ganges, the Irrawaddy, the Mekong, and the Yangtze(10).

It is all a question of Thermodynamics. The only other thing I ever learned from that esoteric science besides those aspects of it which pertain to monsoons and melted babbitt bearings, is that in a closed system Energy tends towards Entropy, and Order tends towards Chaos – and not the other way around. This is expressed in the Second Law of Thermodynamics.

There are theological implications in the Second Law if one considers the Universe as being an "energized clock" that is "winding down" towards entropy, for then one could naturally ask how it got "wound up" in the first place – or Who might have wound it up? If one considers the Universe to be a closed system of infinite proportions, and infinity as cyclic and not linear, then one can run circles around the skeptics until the cows come home.

The Second Law of Thermodynamics seems to have gotten hold of my ordered career in Vietnam, for the last three months of my tour of Duty was nothing but a chaotic juxtaposition of incongruities. I don't know what the theological implication of that was – if any - but perhaps I was being prepared for a Revelation. I know I was soon to receive one, although it was of a decidedly temporal, and not spiritual, nature.

Fire Base Lobo was strictly a dry season habitation and the monsoon rains had begun to flood us out of our bunkers there. Claymore Corners was closer to Phouc Vinh than it was to Phu Loi, so one day - soon after the hoopla of "Operation Twenty-Five Clicks" was over with - Charlie and Delta Company pulled out of the fire base for the last time and moved up to Phouc Vinh.

We were quartered in a little compound of barracks and given a place to park the road equipment. Once again the men could sleep in a hooch and eat in a mess hall. As we had done in the beginning, we ran the equipment down the road to the project every day and brought it back up to the base camp every evening. As we did not have enough lowboy trailers to haul all of the equipment at once, we would have to cut short some of the work early in the evenings to allow for two trips to get everything transported back to base. It was necessary to get everything off the road before dark; we owned the road during the day, but the Viet Cong still had claims on it at night.

One evening soon after we had moved up to Phouc Vinh, Lt. Casazza told me he had to go back down the road on the second lowboy run to pick up another bulldozer. There had been some problem or other and it was getting a little later than usual, and he wanted to get down there and back before darkness caught him. Other equipment was coming in, Dennis had plenty to do, and I told him I would run down there after it. I rounded up a bunch of desperadoes from the Road Gang that had already come in. We mounted a man to ride "shotgun" in the lowboy, crammed the rest into the back of my jeep, threw our beer ration into the floorboard, and off we went down the road.

We found the bulldozer on 1A just above Claymore Corners. I don't remember after all these years exactly what the problem had been – maybe a problem with the truck that had been hauling it - but in any case we got the bulldozer loaded and we all headed back up the road.

We had not gone very far when we met a lowboy from some other outfit coming down the road. It was limping along in the late evening with two blown tires on the trailer, which was carrying a front-end loader. The worried driver stopped and flagged us down. He had gotten a flat on the trailer, which had increased the load distribution on the other tires, causing them to start blowing out. He was rapidly getting into a tight spot, because the next one to go would leave him stranded out there by himself. Could we help?

We told him our old fire base was just down the road a ways, and some itinerant outfit had been occupying it for the last couple of days. We could get him and his loader down there for the night if his truck could still roll on its own. He said if we could haul the loader, his truck with the empty trailer could make it on the tires that he had left.

We put our riflemen out for security while we unloaded our bulldozer and put his bucket loader aboard. I sent the two trucks and a couple of shotgun riders down the road while I remained with the riflemen guarding the bulldozer. That driver was one happy camper, I'm sure, that night down there at good old Fire Base Lobo.

After our lowboy got back we loaded the bulldozer and started again for Phouc Vinh. It was getting pretty late in the evening, and we knew that in the tropics it would get dark quickly. With fifteen or twenty kilometers to go we were bound to get caught.

About three clicks up 1A above Claymore Corners the road dips across rice paddy and then rises to a village on a knoll. Beyond the village, the road passes through more undulations before making its descent into the Song Be River bottom towards the bridge. As our little two-vehicle convoy approached the village, I saw a man standing up there beside the road watching us. As we got closer I could see that he was a big man – bigger than any Vietnamese that I had seen before. He was standing there in the twilight wearing a conical peasant's hat tied under his chin, a brown short sleeved shirt unbuttoned down the front, and nondescript short pants. His sinewy arms hung motionless at his sides. As he stood there bare legged and bare footed he looked for all the world as if he had been rooted to the soil and raising rice for the last ten thousand years – and he probably had been.

I didn't know who he was or what he was doing, but he did not impress me as being a welcoming committee sent out to greet us from the Chamber of Commerce. I cocked my .45 and stuck the muzzle against a sandbag on the floorboard between my feet, and our gallant musketeers looked to their primings.

As we passed by he did not move, but just stood there looking at us. And for a brief moment that I will never forget, the brash, impatient, raucous and parvenu West looked directly into the gaze of the inscrutable East. We were the energy of the bright and lusty cities clamoring day and night for more steam, more steel, more trade, more wealth, more power. He was the eternal blade of grass that pushes up through the cracked-pavement entropy of mean streets and litter-strewn alleyways, of desolate factories and drug-plagued ghettoes, of gang wars waged in crumbling concrete wastelands – the patient blade of grass that waits like the grave for its victory. As we passed by, he looked me in the eye, and he did not blink.

After a few more kilometers we came to the grade that dropped down to the river. The ARVN guard had strung concertina wire across the road for the night. I did not see it in the gloom of dusk until we were about halfway down the hill. The guards saw us coming and looked as though they were going to challenge us before permitting us to pass. We were coming down the hill at a fairly good clip and I started thinking about that truck and that bulldozer breathing down our necks. We started blowing the horn for the guards to get the wire out of the way, and they evidently perceived what was going on and started scrambling to clear the roadblock. The truck was coming down grade like the Old 97, and I could hear his air brakes moaning behind us. When we got to the bottom I told the jeep driver to pull over quick and get out of the way. As soon as we did the rig roared by and ran through the roadblock, scattering jabbering ARVN soldiers like a flock of chickens. I never saw so much commotion in one spot in all my life since that eel truck turned over down in Gloucester County.

When the truck finally got stopped, there was a wad of wire wrapped around the radiator cowling and jammed up under the chassis, and long coils of concertina wire strung down the sides like catfish whiskers. By the time we got untangled from all that it had really gotten dark. We mounted our dauntless convoy once again and headed north across the Song Be Bridge, with ten more clicks to go.

Riding up 1A that night I felt like I might as well be marching up the Ho Chi Minh Trail with the Preservation Hall Jazz Band, but we finally made it to the gate at Phouc Vinh without further incident. I was glad to see it. It was closed at that hour but an MP stepped out of the guardhouse and asked us where we were going. I told him. I guess he was not used to traffic coming to his gate out of the blackness of night. He went back into the guardhouse and made a call on the land line. He came back out and said he had to clear something with somebody and was waiting for his call to get through. I said did we look like Viet Cong infiltrators? No, Sir, we certainly did not, but he just had to clear this with his superior. He was not supposed to open the gate after dark.

I was probably getting a little impatient with our adventure by then, but I could just imagine us sitting out there all night long waiting for some red tape to unravel. Tiny, our great-sized and great-hearted Tennessean, was one of the desperadoes we had with us in the back of the jeep. I turned and asked Tiny if he would mind explaining to this gentleman who we were. Tiny said that he would be delighted to do so.

Tiny did not have on a shirt at the time, and as he rose up out of the back seat of the jeep he looked just like some big Tennessee bear that had come rolling down out of the Cumberland Mountains. He told – or sort of roared at, really - the MP that we were from "Dusty Delta's" Road Gang, 34th Engineers, and that if he didn't open that (expletive deleted) gate in about two seconds we were coming through it anyway, and that it wouldn't be the first one we'd gone through like that tonight, either!

The MP evidently considered that to be adequate authorization for him to open the gate, and he waved us on through. By the time we got back to the Company area with that bulldozer I was about ready for another beer.

We were putting the finishing touches on the road project by the first of May. The rains had come but they were still intermittent enough to allow work to continue. Lt. Meislin was finishing his last culvert, and Lt. Casazza was on the home stretch with the dirt. It seemed that the pressure was off of us at last. I drove back down to Phu Loi to check on things there and to report to the Colonel.

When I got there I went to the showers and cleaned off the road dust. While I was standing before the mirror combing my hair, the S-3 came in and told me to trim my mustache, the Colonel wanted to see me. When I reported to the Colonel, he told me that I was relieved of my command.

I felt like I had been punched in the solar plexus. What was the reason?

Lack of maturity, I was told.

I thought back over the last few months. I thought about my Christmas "improvement of orders," and my disobedience the day of the helicopter rides; the monikers painted on the equipment, and the Ace of Spades; my lack of communication about the material calculations – twice – and the episode at my briefing of the Brigade Commander in front of the Colonel. I thought especially of that.

I had never gone to the Colonel afterwards to explain that I had called down the night before to try and tell him I had re-figured the schedule. I, in my innocence, had assumed that the S-3 had told him like he said he would, and I couldn't understand why the colonel had been shocked at my briefing of the Brigade Commander on the road that day. I hadn't considered until it was too late that the S-3 had not told him. And then I thought of the payback I got for it with my two men and the medals they didn't receive at the ceremony on the pavement end. I thought that was pretty crummy.

But that was only the capstone to the continuous undercurrent of pressure we received for our being behind on S-3's mis-calculated schedule, while at the same time we were being compelled to make "bricks without straw" with our lack of critical equipment. I could understand that the problem with the scrapers was beyond the control of the battalion, but the little issue with the sheep's foot roller was not. For the XO to not let me go get that so we could get the job done seemed petty and vindictive to me. It caused me to start questioning motives in some quarters. And when I saw the pavement going down on a monsoon-soaked base course for "Operation Twenty-Five Clicks," only to have the surface fall apart after somebody got a star, answers to some of my questions seemed to reveal themselves like an Epiphany. This Revelation was then extended beyond our road project and expanded to encompass the whole war.

After that I admit I started feeling bitter and cynical, and it showed up in some erratic behavior. I'm afraid my "Cadet Private Syndrome" had reared its ugly head again – the one that had kept me on barracks confinement and marching penalty tours at VMI for so long. I suppose it all added up to a pretty big indictment against me. Yes, indeed, lack of maturity.

Well, I don't suppose I could have argued with that, but – Revelation or not - I was so ashamed. What would my Father think? What would all of my Uncles think, and my Forefathers who had fought in the wars all the way back to the Revolution? Theirs was not to reason why - they were Gentlemen, Disinterested and Honorable. I had brought shame on our family name – and, as a VMI man, I had not added any lustre to VMI's name either. I wanted to go back up in the Blue Ridge Mountains somewhere and crawl into a hollow log like some wild animal looking for a place to die.

I didn't know what to do. I thought about getting a transfer to some Land Clearing outfit. That would be a good place for me – out in the woods with the rest of the misfits. But then I thought that I would only be running away, and it wouldn't change anything. I could run away from the 34th Engineers, but I couldn't run away from myself. The only thing to do would be to stand up to my shame, and try to redeem myself with the battalion in some way. I had three more months to go. Maybe there would be time.

My first Duty under the new order of things was to help my replacement make the transition. My relief was Capt. Dick Daugherity, VMI Class of '66, and I took him up to the road and showed him around. I told him he had a good company, both down at Phu Loi and up on the road, and that he could depend on all of them.

After turning over things to Capt. Daugherity, I was sort of a floater. Whenever a directive came down for the Battalion to provide the name of an Officer for some detached Duty, mine was the name that was submitted.

My name was first submitted for the position of Aide to Major General Dillard, Commander of Engineer Troops in Vietnam. It would have been an honor to serve him, and I thought that there was a good chance that I would be selected. General Dillard, VMI Class of '42, had remembered me whenever he had made his inspections on our road project, and I have always felt that he remembered me in this instance. I sense now that he must have considered the position as being not the best for the career track of a Regular Officer, and that I would be better off remaining in a line unit dealing with troops, gaining leadership experience, and making command decisions. I know, without ever having been told, it was because of his thoughtful consideration that my life was spared. A week later, on 12 May, 1970, his helicopter was shot down near Pleiku and all aboard were killed. I would have been with him.

Major General John Dillard, VMI '42 - one of the "Old Corps" - was the senior ranking American Officer to be killed in Vietnam. He was one of the last veterans of World War II still on active Duty. It was the end of an American era. With the death of the Fisher King, the Grail was lost and the land withered. Where now is to be found the questing knight who can find the Chapel Perilous, and who knows the question to ask there that will restore us?

None here in this wasted land, where the Chapel is only the wind's home....

Some weeks thereafter, the Viet Cong struck C Company's road site, and struck it hard. They knocked out seven pieces of equipment with rocket propelled grenades, killing three or four men and wounding a number of others. I was told that afterwards an extensive network of entrenchment was found that had provided the enemy close access to C Company's project site, indicating that elaborate preparations for making the attack had been going on for some time prior.

It had occurred, I believe, in the vicinity of that rice paddy that drained into one branch of the swamp east of Claymore Corners, near that last place where we had looked for laterite. The rice paddy was a few clicks above Claymore Corners, just south of the village. I immediately thought of that man we had seen standing there by the road watching us go by with the bulldozer that evening, and was then convinced that he must have been a Communist Chinese adviser to the local Viet Cong. I thought at the time that he looked too big to be Vietnamese.

Just as Charlie made his attack, the Colonel and the Major were topping the rise in the Colonel's jeep, and they had to dive for cover in the ditch. It was a sad lesson, but it vindicated me on the security precautions I had taken up there on the road - if there had ever been any question.

Prior to the attack on C Company, and about the time of General Dillard's death, I received word that some of the men up at Phouc Vinh wanted to see me, and hoped I would come up the following afternoon. Of course I would. Somebody provided me with a jeep and driver and I arrived at the requested time. There I found that Sergeant Mendoza had gotten together a group from the Road Gang. There was a big cooler of beer beside a picnic table, and I was touched by their thoughtfulness. We popped a few cold ones together and chatted awhile, and I asked about the road and how everyone was doing. It was good to see them again.

Presently Sergeant Mendoza said they had a little something for me to remember them by, and he presented me with a plaque. It was a piece of dark lacquered wood in the shape of a shield. At top center was a scroll that said Vietnam, flanked by the American flag on the left, and the flag of the Republic of Vietnam on the right. Centered below the scroll was a likeness of a map of North and South Vietnam. It was flanked on the left by the Engineer Castle, and on the right by a likeness of our shoulder patch – the emblem of the 20th Engineer Brigade. And centered at the bottom was a silver plate, with the inscription:

Presented to

Cpt H.V.Traywick "Bo"

To the best damn C.O.

D Company ever had

"All The Way Sir"

Proud to serve with you

"Anytime – Anywhere"

From Dusty Delta's ROAD GANG

D Company 34th Engr. Bn

Phu-Loi, F.S.B.Lobo, Phouc-Vinh

While I was looking at it, a bit of laterite dust or something blew into my eye, making it hard for me to read.

Still is.....

I never saw them again.

Driving back down to Phu Loi, we took the jeep around the bypass road past old Fire Base Lobo. It was empty and desolate. The rains had already begun to take effect upon the place, and the unattended bunkers had started to crumble. The fire base looked sad and forlorn, and we drove on. I never saw that place again, either, except in the fragments of dreams.

Air Strike at the Water Point
Viewed from the Parapet of Fire Base Lobo

CHAPTER IX

THE WASTE LAND

The rest of my tour was like a travelogue. I was still floating around battalion, with my name still being submitted for every slot that came down the pike. My next Duty was to be liaison for a USO troupe. Talk show host Johnny Grant was coming over from California with a Playboy Bunny and another girl to go around and inspire the troops. I was to go with them and serve as their Liaison Officer for a week or ten days.

We were based out of Saigon, but I saw more of South Vietnam in that period than I did in all the rest of my year's tour of Duty. We flew about in Huey helicopters: from III Corps up to II Corps, then down to the Delta and back again. We flew by courier jet out to Yankee Station in the South China Sea and landed on a Carrier to visit the Navy. We met the Admiral, and the girls met many of the sailors before we were transferred to a Destroyer by helicopter. Those "tin cans" are livelier than a Carrier, and some of the troupe got a little seasick, but they carried on bravely with their morale mission as we watched an Oiler come alongside and refuel our ship.

Back ashore we sat in the warm sun on the beach at Qui Nhon, then felt the chill in the air in the Central Highlands at Pleiku with the 4th Infantry Division – my Father's old outfit before he went to the Far East in World War II. Down in the Delta we saw a sea of green stretching to the horizon in all directions under a vast azure sky, broken only by a thin row of rickety brown huts perched on stilts beside a canal – a picture that lingers in the mind's eye like a vision from a surreal dream. We dined in style in Saigon hotels in between excursions, and once had lunch with General Creighton Abrams, Commander in Chief in Vietnam. It is not every day that one has lunch with a four-star General and a Playboy Bunny at the far end of the earth.

But these were only momentary interludes between visits to troop units in the field. We flew countless miles looking out of the open door of a Huey at the tropical mosaic passing below us like a lazy kaleidoscope. Sometimes after take-off the chopper-jockey would hot-dog it up a ravine, banking right and left as the greenery rushed by, to give the girls a thrill before we gained altitude and headed for the next destination.

Our helicopter would set down in some God-forsaken place on the border, with a washed out monsoon sky overhead and a soggy, slippery red dirt underfoot. Among sandbagged bunkers, and pallets of ammunition and supplies strewn about the nondescript jumble of a fire base at the far end of nowhere, an olive-drab, rain-soaked group of men would be waiting to see this apparition come down from the sky. Mr. Grant and I would stay back by the chopper while the girls would go over and talk to the men crowding around. If there were winks and nudges and whistles and cat-calls, I never saw or heard the first one. This was not World War II.

There seemed to be an almost reverent quietness about those men. These were not Playboy Bunnies and California "chicks" who had come out to see them; these were those men's girls from back in "The World" – those lucky ones whose wives or whose lovers had not forsaken them by now. It was like a representation of all the people left in the world who - if they were really lucky they could count on the fingers of more than one hand - gave a damn about them, and who had come out there to see them.

And I saw them file down out of the jungle off some patrol, and saw them standing in the mud - all sweat-stained and awe-struck as no Viet Cong could make them awe-struck - and I saw them crowd around, in a dark drab silent mass, these two bright girls who shone in their midst with an aura about them like Mary Magdalene and the Holy Virgin. The shimmering illusion that they gave to those men was the last of home that some would ever see. Heaven holds a place for those two young Ladies.

After my return to Phu Loi, I figured I might as well take my (legal) R&R and get it over with. I went with a Lieutenant from the office of the S-3 who lived in a hooch next to mine. He had an easy manner and a ready charm that women can't resist. He had been a ski bum in Europe, and evidently his charm had gotten him into a situation in which the judge had given him two choices, one of which was to join the Army and go to Vietnam.

We decided to go to Bangkok, Thailand. I was afraid that was going to be nothing more than "Saigon West," but that turned out not to be the case. The Thai have the delicate bone structure of the Vietnamese, but we found that they didn't have the angular gauntness. Perhaps that is the difference between peace and war. In addition we found that the Thai - and the Malay - sometimes have an infusion of European blood that makes for some of the most exotic and beautiful women in the world.

When we landed at the airport, a tailor came up to me and took my measure, and the next day delivered to me a double-breasted, pin-striped suit that fit perfectly. I wore it for years after I got back to the States before finally giving it to my friend "Blackie" Abbott, who lived in the back room of Geezer's place out towards Folly Beach, South Carolina.

While in Bangkok we saw the sights about the city. We saw a seventy-foot long reclining Buddha, and a six-ton Buddha made of solid gold. We saw gold shops in slums next door to opium dens. We smelled the cooking of exotic foods at shops and stalls along streets choked with vehicles of every description blowing their horns like a traffic jam of Bedlamites. And we saw the elephants piling teak, in a "sludgy, squdgy creek," just as Kipling had described them(11).

At night we heard the East a-callin' and we answered the call. The discos were wide open. Thai bands would play American rock - singing words learned phonetically - and they gave some pretty good renditions. Dance partners could be had for the price of dinner and a taxi ride home, and by the time that week was over with, we had both come down with a pretty bad case of entropy.

When we got back from R&R, I found out that the Battalion was to be sent to the Delta in about three months - after all the paving had been completed on the Highway 1A/2A Upgrade Project. The next mission of the 34th was to be the upgrade of the road that ran from Can Tho up along the southern bank of the Mekong River to the Cambodian border. Someone had to be sent down there on "Engineer Reconnaissance" to make preparations. I requested to be the one sent, and the Colonel agreed – probably glad to be rid of me at last. It promised to be an interesting assignment: it would get me out of Phu Loi and away from the flagpole; it would occupy me for the rest of my tour of Duty; and, most especially, it would give me the opportunity that I had hoped for to redeem my Honor.

A base camp area had to be prepared for the battalion before it moved down there. Land Use Concurrences (LUCs) for the road right-of-way must be obtained from the local Vietnamese Province officials. A rock quarry had to be established on a mountain near the Cambodian border to provide the road-building material, and a canal had to be laid out to barge the rock from the quarry over to the Mekong River. A civilian contractor from the States - Pacific Architects and Engineers (PA&E) - was to do the dredging for the canal. They were also to pump dredge spoil onto the marsh where our base was to be located, in order to build up a piece of solid ground upon which the battalion could set up camp. Before the spoil could be pumped, a dike had to be laid out and built to contain it.

My assignment was not to begin any actual construction, but to do the surveys and lay-out work, and to co-ordinate with the local officials and with PA&E. I was to return to Phu Loi periodically and report to the Colonel on the progress of the preparations.

It was near the end of May when I packed my kit and headed for the Mekong. I had to hitch-hike on helicopters. Luckily there were plenty of them around Phu Loi, so departure was never much of a problem. Arrival in the Delta, however, was sometimes problematic. Often I would just have to take a chopper that was going in a southerly direction, and see what else I could catch wherever I was put down. One time I was left for the night somewhere along the way and was served a lobster dinner at some Officer's mess. Another time I was left in a chicken lot in My Tho, half way down to Binh Thuy, with the sun going down and nothing and nobody around but a land line telephone over in a shed. I finally got hold of somebody and a passing chopper stopped by and gave me a lift to my destination on the river.

The Mekong is one of the great rivers of the world. Its origin is in the Tibetan Highlands of western China, and it flows southeastward for 2,600 miles down through the Indochinese Peninsula before reaching the

South China Sea. The river forms the border between Burma and Laos - and then the border between Thailand and Laos - before flowing into Cambodia. Below the Cambodian capital of Phnom Penh it branches into its delta just before entering Vietnam. The southern branch of the delta enters Vietnam at Chau Doc. About seventy kilometers down river from that place is Binh Thuy, where was located the Headquarters of the 20th Engineer Brigade.

I was given a billet, a desk at Brigade Headquarters, and a jeep whenever I needed one. I often went up the road, which went as straight as a die along the river from Binh Thuy to the Cambodian border. About half way up the road was Long Xuyen, near where our proposed base camp was to be set up. There I would confer with Province officials and try to get permission to use the land – which in reality was just marsh. I also had to get LUCs for the road right-of-way. I had to use an interpreter, and trying to deal with a bureaucrat in a foreign language is an experience that is quite challenging. If you are required to get any results it is even more so.

Dealing with the American civilian contractor was almost as much of a challenge. Some said PA&E stood for "Procrastination, Alibis, & Excuses," but to be fair about it, the Viet Cong had recently blown up one of their dredges, so that must have had an adverse impact on their scheduling.

Our rock quarry was to be established near Chau Doc. Just to the southwest was the outcropping of some mountains in Cambodia that spilled over into Vietnam. These were known as the Seven Sisters. The one closest to the river was Nui Sam, and on this was to be established the quarry. There was a road running six kilometers down a causeway across the marsh that connected Chau Doc with the mountain. On the left side of the causeway was Vietnam, and on the right was Cambodia.

It was such a long way up the road from Binh Thuy that I often stayed at Chau Doc at the MACV compound there. I was given a survey crew from Brigade, and we ran extensive surveys for the quarry and canal. We took transits and chains and ran traverses all over the mountain, locating ARVN compounds, cemeteries, shrines, and all the other things that had to be avoided when the quarry was opened. We also had to lay out the canal, and we took stadia rods out into the vast marsh and surveyed the route from the mountain to the river. Then we took the data back to Binh Thuy and made plats and drawings of everything.

The Vietnamization of the war was probably as much in evidence in the Delta as anywhere. The 9th

Infantry Division had been the big presence of Regular U.S. troops in the region, and the departure for the States of all but one brigade left a vacuum that the North Vietnamese Army (NVA) did not hesitate to capitalize upon. They slipped down the Cambodian border and into the Seven Sisters, where they built an elaborate cave complex in Nui Sap, in the heart of the Mountains. I understand they even had a hospital in the caves. B-52 strikes would be made against them there, and the awesome bombing would sometimes catch them in the open and atomize them, while the ones inside the caves would have their ear drums ruptured by the concussions.

I met a MACV Lieutenant at Chau Doc who had been stationed at an ARVN fire base down in the Seven Sisters. Above the fire base was a slope that was covered by scrub brush. Time and again the MACV Advisers implored the ARVN Commander to send his men up there and clear off the brush. The ARVN Commander kept putting them off, saying that he had requested defoliation (Agent Orange). This went on for two months with no results, until one night the NVA – concealed by the brush - slipped down the slope and overran the fire base. They threw satchel charges into the bunkers, killing most of the MACV Advisers and many of the South Vietnamese soldiers. My Lieutenant friend was there. He said they found what was left of the MACV Captain splattered all over the inside of his bunker. This

Lieutenant said he had finally gotten disgusted with the ARVN and had asked for reassignment, and now he was being ostracized for it.

One day during my USO Liaison we put down at a land clearing operation somewhere in the Delta. While we were there, I spoke with a bulldozer operator who had a hole in the blade of his Rome Plow made by an RPG, and a dent in the roof of his cab made by a mortar round. I asked him to tell me a war story. He said that the ARVN were supposed to have been providing them with security, but when the Viet Cong attacked they found that they had been abandoned, and they had to fight off "Mr. Charles" from their bulldozers with their M-3 submachine guns.

Many times riding up the road I would see an American truck tearing recklessly by with an ARVN at the wheel, and I have seen a number of them upside down in the marsh or rice paddy. One night all the lights went out in our area at Binh Thuy. We found out later that the switchboard panel over at our generator point had started smoking, and the Vietnamese attendant had solved the problem by throwing a bucket of water onto it. Moving right along with the Vietnamization process.....

Somewhere along the line, the Colonel, the Major, and others of the old crowd at the 34th got "short" and then went back home. One by one, we had new faces around the battalion, and I hit it off famously with the new Battalion Commander. He came down to the Delta, and I showed him around and filled him in on my progress. He was evidently favorably impressed – and I was evidently learning to communicate.

I, too, was getting to be a short timer – "so short I could sit on a dime and swing my legs off the sides;" "so short I had to stand on tiptoe to look under the table" – and then one day I got my orders home. A few days before my scheduled departure I left the Delta for the last time, and came back to Phu Loi to wrap things up on my progress report to the new Colonel.

That night there was a little ceremony held in the Battalion Officers Club, at which I was awarded the Bronze Star for "meritorious service in connection with military operations against a hostile force in the Republic of Vietnam." That same night I was told that the battalion was not going to be moving to the Delta after all. But I did not consider my efforts of the past three months to be an exercise in futility, for it had allowed me the opportunity to redeem my Honor. I felt that the Bronze Star had been given to me largely because of those efforts, and I considered that recognition to be the measure of my redemption with the 34th Engineer Battalion.

* * *

My last night in Vietnam I was sitting at the bar in the club. Everyone else had turned in for the night. I was talking to the bartender and listening to The Byrds' "Ballad of Easy Rider" on the tape deck behind the bar. Just then there was a BOOM, and clods of something started raining down onto the tin roof of the building. I was just plain too short to be worrying about all that any more, so I bought the bartender and me another drink and we went and sat under the bar. About that time our "Fightin' Parson" came charging into the club in full battle array – helmet, flak jacket, and M-16 – shouting "Incoming! Incoming!" I poked my head up from behind the bar and said "Chaplain, won't you join us in a little libation?" Before he could reply, Lieutenant Mike Kinh, good old "Dusty Delta" Construction Platoon Leader, came in and told us it was OK - somebody had just fragged the NCO's latrine again. Nobody was in it at the time.

And that's the way it was - August of 1970, Republic of Vietnam.....

The next day I went around and told everyone goodbye, and wished them good luck. The crusty old

Battalion Sergeant-Major – who had probably been in the Army since the Spanish-American War – told me as I was shaking his hand: "Sir, it was the Major that railroaded you." I thanked him for telling me. I didn't know anyone had railroaded me – I thought I had succeeded in jumping the track on my own. I had become rather accomplished at that as a Cadet at VMI. But that cleared up some things – sort of.

I don't know what the Major had against me. Maybe he was the one who had lost face about the incident over the melted babbitt bearings in the oil pan of that scraper. I would have thought that if anyone would have a grudge against me it would have been the S-3 over the material calculations, but I don't see what the Major's involvement would be with that - unless he heard about it over a pinochle game and just wanted to take out his personal problems on somebody handy. I never did think that the Colonel came up with the snub of my two men at the awards ceremony up on the road after "Operation Twenty-Five Clicks." Somebody proposed it to him and convinced him to go with the idea - and it might have been the Major. I don't know, and speculation now does no-one any good. Maybe there was nothing personal. Maybe things were just getting to everybody by then, like they were to me. I was certainly no shining example of equanimity.

But that is what the Sergeant-Major told me. I admit I had a lot coming, even if I didn't deserve all that I got. Forget it. The Major was long gone, and subsequent events had vindicated me on all the important things: our security precautions, the material calculations, and the Project Schedule. The Road Gang had built a solid section of road as a monument to their Duty well done. The good Colonel – before he left - had given me the opportunity to redeem my Honor, and I felt I had done so. I still had all of my body parts, and in another couple of hours I would be boarding the "Freedom Bird" back to the World. Things had turned out all right after all.

As luck would have it, I flew back to the States with Tom Brannen, from Brazil, Indiana. Tom had driven the mail truck for D Company, and we would always be glad to see the dust from his three-quarter ton coming up the road to the fire base. We sat together, of course, on the flight home.

I remember the great cheer that went up in the cabin of the plane when it lifted off from the airstrip. My landing there a year ago seemed like a dream. We flew the Great Circle route to New Jersey, refueling in Tokyo and in Anchorage. I remember the chill in the air that night in Alaska as we stepped out onto the tarmac to stretch our legs.

Arriving in New Jersey the next day, Tom and I decided that the transition was too abrupt, and that we had better "decompress" a little bit before we headed home. We got out to the Jersey Shore somehow, checked into a room over a beach bar, and then went downstairs to condition ourselves for this new way of life. But first I called the folks from a pay phone in the bar. My Mother answered:

"Hello?"

"Hi, Ma! I'm back!"

She just started crying.

I remember the bartender asking to see my I.D. I guess Tom and I didn't quite fit the profile of the crowd - but we could make allowances for that, for we had come from the wilderness and had been living on locusts and wild honey for many a day.

The next morning the sun was streaming through the open window of our room in a most troublesome fashion. The first thing I heard was the raucous voice of some New Jersey woman yelling at somebody under our window: "AHH! GETCHA GOID DIMN DOIG OFFA THE SOIDWALK, YA BUM YA!"

Not home quite yet, I said to myself.

We caught the bus south. At Philadelphia, Tom and I parted company – he for Indiana, and me for Old Virginny. We had come a long way together – across a Great Divide. As my bus pulled out I saw a gang of Afro-adorned Black Militants standing outside the bus station – Aquarian Pilgrims come up here looking for the Promised Land in the City of Brotherly Love. They were the meanest looking bunch of people I have ever seen. Something in the Promised Land had rotted their souls.

Still a long way from home.

It had begun to drizzle when we came into Washington that evening. John Winthrop's "City Upon a Hill" had long years before relocated here and gone secular, but it was still working the Crusade hustle after all these years – the latest angles being Democracy for the Asians and a Brave New World for us.

Already had some, thanks. Help yourself. Charity begins at home, and the only Crusade I'm interested in is getting myself over yonder to the south side of that river.

We crossed the Potomac into Virginia and I could tell it in my bones. The bus rolled on down Route 29 along the eastern foothills of the Blue Ridge Mountains, past names that seemed to come to me from a land lost somewhere in a mist-shrouded reverie: Fairfax, Manassas, Warrenton, Culpeper, Charlottesville, Monroe.

At last we dropped down the hill from Madison Heights and crossed the old James River, reaching the bus station at Lynchburg late in the night. I got a taxi to carry me out to the house. I remember hearing the gravel crunching under the tires as we pulled around the driveway. In the darkness I looked up and saw a large banner, somewhat bedraggled from the rain, strung across the front of the house:

WELCOME HOME OUR HERO

Home! For some, that had become a place that would be difficult to find.

Convoy on D Company's Road Section
Highway 2A

EPILOGUE

~

"SHIP ME SOMEWHERES EAST OF SUEZ….." (12)

All that was a long time ago, and the road winds on. Most all of the veterans of those times are civilians now, and have been for over forty years. Many have grandchildren who study Vietnam in school, and consider it as much a part of ancient history as the Trojan War. I don't know what happened to the men of the old Road Gang. I hope they are all well and prosperous, with happy families and grandchildren tumbling around their knees. They deserve it all – for they are as good as any men who ever picked up a rifle in defense of their country, or wrestled a bulldozer in a monsoon rain.

I sometimes think of how I would like to have gone back to the Far East with my Father. I would have gotten that tough old jungle fighter to show me where they cut the road at Maingkwan and came near to destroying the Japanese 18th Division. And I'd like to see the airstrip at Myitkina, and the railhead at Ledo, and the Ledo Road – "Pick's Pike," built by General Pick, from down there in Brookneal, in the good old red clay of Southside Virginia, not far from where Patrick Henry is buried. Then we could have gone to Vietnam together and I would have shown Dad the road section, and maybe seen what was left of Fire Base Lobo. There would be something left, I know, for the archaeologists say that the hardest thing in the world to get rid of is a hole in the ground.

But the fire base had already started to crumble when I saw it last, and it has no doubt by now been overgrown by wild rubber plantation and scrub underbrush. The bypass road would have been reclaimed as well, and one would have to thrash around in the woods to find anything - like some Mayan ruin in the jungles of Yucatan. And in some issue of National Geographic there were pictures showing that the jungle had also reclaimed the Ledo Road in Burma, leaving little but ruins that my Father could have found of his service there either.

So maybe it is better not to go back to the East that way, for the only thing that awaits us there is a stark confrontation with mortality. It is better, I think, to go back in memory, where everything still abides.

And I do go back sometimes, when a summer downpour thunders on the tin roof of a screen porch, or when the breeze stirs a certain set of wind chimes, or when a fragment of chatter from the kitchen of an Oriental restaurant comes to my ears. Then I will see again *papa-san* going by on his mo-ped with his lovely daughter riding side-saddle on the back – her colorful *ao-dai* fluttering in the breeze like a butterfly dancing in the sunlight; and I will feel again the relentless tropical sun beating down on my shoulders; and I will hear again the clacking of the palm leaves in the riotous greenery of some of the most beautiful places on earth. And once again I will be up there with the Road Gang, amid the dust and the roar of the diesels on Highway Two Alpha.

Back then we were very careful to always refer to each other as men, but now I guess we are all old enough to not mind being called boys. And wasn't that the best time after all, when we were young and immortal up there on the road, finding out what kind of stuff we were made of, and living for something bigger than ourselves?

Can't you hear it now, old boys, wherever you might be? The roar of the diesel engines in the laterite pit, the thumping of the pogo sticks on the culvert site, the rattle of the machine gun down at the water point? It's all there, whenever you want it: the water truck spraying the road; the sheep's foot roller patiently plodding along behind a clanking bulldozer; the grader cutting the finest of a fine shave onto the crown; the lumbering pans coming down the road with another load of dirt. And hanging over all is a thick red cloud of dust, rising in the tropical heat like smoke from the pyres of our innocence and drifting slowly away on the wind....

END NOTES

Title Page:

1. Siegfried Sassoon, *Counter-Attack and Other Poems* (New York: E. P. Dutton & Co., 1918) 10.

Chapter III:

2. Stanley Karnow, *Vietnam: A History*, Second Revised and Updated ed. (New York: Penguin, 1997) 134-8.

3. Thomas Carlyle, "Latter-Day Pamphlets No. IV – The New Downing Street" in *The Works of Thomas Carlyle*, 12 vols. Library ed. (New York: John B. Alden, 1885) 8: 133-4.

4. Rudyard Kipling, "Gunga Din," *Rudyard Kipling's Verse*, Inclusive ed. 1885-1918 (Garden City & New York: Doubleday, Page & Co., 1919) 462.

5. William Cabell Bruce, *John Randolph of Roanoke 1773-1833*, 2 vols. (New York & London: G. P. Putnam's Sons, 1922) I: 418-9.

6. Polybius. *The Histories*, Trans. Robin Waterfield. Oxford World Classics Series. (Oxford: Oxford UP, 2010) 372-3.

7. Hannah Arendt, *The Origins of Totalitarianism*, First Expanded ed. (Cleveland & New York: Meridian – World Publishing Co., 1958) 322-4.

8. Robert H. Bork, *Slouching Towards Gomorrah: Modern Liberalism and American Decline* (New York: ReganBooks – HarperCollins, 1996) 331-2.

9. Bruce, *Randolph* I: 418-9.

Chapter VIII:

10. H. O. Pub. No. 9. *American Practical Navigator: An Epitome of Navigation*, originally by Nathaniel Bowditch, LLD (Washington: U. S. Government Printing Office, 1966) 799-800.

Chapter IX:

11. Kipling, "Mandalay," *Verse,* 478.

Epilogue:

12. *Ibid.* 478.

SOURCES

Arendt, Hannah. *The Origins of Totalitarianism*, First Expanded ed. Cleveland & New York: Meridian – World Publishing Co., 1958.

Bork, Robert H. *Slouching Towards Gomorrah: Modern Liberalism and American Decline*. New York: ReganBooks – HarperCollins, 1996.

Bruce, William Cabell. *John Randolph of Roanoke 1773 – 1833*, 2 vols. New York & London: G. P. Putnam's Sons, 1922. Volume 1.

Carlyle, Thomas. "Latter-Day Pamphlets No. IV – The New Downing Street," in *The Works of Thomas Carlyle*, 12 vols. Library ed. New York: John B. Alden, 1885. Volume 8.

H. O. Pub. No. 9. *American Practical Navigator: An Epitome of Navigation*, originally published by Nathaniel Bowditch, LLD. Washington: U. S. Government Printing Office, 1966.

Karnow, Stanley. *Vietnam: A History*, Second Revised and Updated ed. New York: Penguin, 1997.

Kipling, Rudyard. *Rudyard Kipling's Verse*, Inclusive ed. 1885-1918. Garden City & New York: Doubleday, Page & Co., 1919.

Polybius. *The Histories*, trans. Robin Waterfield. Oxford World Classics Series. Oxford: Oxford UP, 2010.

Sassoon, Siegfried. *Counter-Attack and Other Poems*. New York: E. P. Dutton & Co., 1918.

ABOUT THE AUTHOR

Author H. V. "Bo" Traywick, Jr., graduated from the Virginia Military Institute in 1967 as a Distinguished Military Graduate (DMG) with a degree in Civil Engineering. He was commissioned in the Regular Army as a Second Lieutenant in the Corps of Engineers and was sent to Ft. Belvoir, Virginia, for Engineer Officer Basic Training. Upon completion, he was sent to Ft. Benning, Georgia, where he successfully completed Ranger training. In 1968, he was assigned to the 575th Engineer Battalion (Construction) at Ft. Stewart, Georgia. His service there included Platoon Leader, Company Executive Officer, and eventually Company Commander.

During this period, he was sent TDY to Ft. Sherman in the Canal Zone, where he successfully completed the Jungle Operations Course. Just prior to his promotion to Captain and his deployment to Vietnam in August of 1969, he successfully completed the Parachute School at Ft. Benning. In Vietnam, he was assigned as Base Development Officer with the 3rd Brigade, 82nd Airborne Division, at Phu Loi, where he was awarded the Army Commendation Medal. Upon the 82nd's redeployment to the continental United States (CONUS), Captain Traywick was assigned to the 34th Engineer Battalion (Construction) at Phu Loi. He was given command of D Company and was engaged primarily on a road construction project on the Main Supply Route (MSR) for the 1st Air Cavalry Division based at Phouc Vinh. His last three months in country were spent in the Mekong Delta on the Cambodian border, near Chau Doc, engaged in Engineer Reconnaissance. Just prior to redeployment to CONUS, he was awarded the Bronze Star. Upon his return, he resigned from the service and made a career as a tugboat captain, retiring in 2010. The author presently lives in Richmond, Virginia, where he reads, writes and serves part time as a tugboat pilot.